Geopiracy

DOI: 10.1057/9781137301758

Other Palgrave Pivot titles

G. Douglas Atkins: T.S. Eliot Materialized: Literal Meaning and Embodied Truth

Martin Barker: Live To Your Local Cinema: The Remarkable Rise of Livecasting

Michael Bennett: Narrating the Past through Theatre: Four Crucial Texts

Arthur Asa Berger: Media, Myth, and Society

Hamid Dabashi: Being a Muslim in the World

David Elliott: Fukushima: Impacts and Implications

Milton J. Esman: The Emerging American Garrison State

Kelly Forrest: Moments, Attachment and Formations of Selfhood: Dancing with Now

Steve Fuller: Preparing for Life in Humanity 2.0

Ioannis N. Grigoriadis: Instilling Religion in Greek and Turkish Nationalism: A "Sacred Synthesis"

Jonathan Hart: Textual Imitation: Making and Seeing in Literature

Akira Iriye: Global and Transnational History: The Past, Present, and Future

Mikael Klintman: Citizen-Consumers and Evolutionary Theory: Reducing Environmental Harm through Our Social Motivation

Helen Jefferson Lenskyj: Gender Politics and the Olympic Industry

Christos Lynteris: The Spirit of Selflessness in Maoist China: Socialist Medicine and the New Man

Ekpen James Omonbude: Cross-border Oil and Gas Pipelines and the Role of the Transit Country: Economics, Challenges, and Solutions

William F. Pinar: Curriculum Studies in the United States: Present Circumstances, Intellectual Histories

Henry Rosemont, Jr.: A Reader's Companion to the Confucian *Analects*

Kazuhiko Togo (*editor*): Japan and Reconciliation in Post-war Asia: The Murayama Statement and Its Implications

Kath Woodward: Sporting Times

DOI: 10.1057/9781137301758

palgrave▸pivot

Geopiracy: Oaxaca, Militant Empiricism, and Geographical Thought

Joel Wainwright

palgrave
macmillan

DOI: 10.1057/9781137301758

GEOPIRACY

First published in 2013 by
PALGRAVE MACMILLAN®
in the United States—a division of St. Martin's Press LLC,
175 Fifth Avenue, New York, NY 10010.

Where this book is distributed in the UK, Europe and the rest of the world,
this is by Palgrave Macmillan, a division of Macmillan Publishers Limited,
registered in England, company number 785998, of Houndmills,
Basingstoke, Hampshire RG21 6XS.

Palgrave Macmillan is the global academic imprint of the above companies
and has companies and representatives throughout the world.

Palgrave® and Macmillan® are registered trademarks in the United States,
the United Kingdom, Europe and other countries.

ISBN: 978-1-137-30174-1 EPUB
ISBN: 978-1-137-30175-8 PDF
ISBN: 978-1-137-30173-4 Hardback

Library of Congress Cataloging-in-Publication Data is available from
the Library of Congress.

A catalogue record of the book is available from the British Library.

First edition: 2013

www.palgrave.com/pivot

DOI: 10.1057/9781137301758

For Inés and Inez

DOI: 10.1057/9781137301758

Geography invites *exploration* [...] emphasizes *location*
[...] involves *measurement.*

I. Bowman (1930)

[M]y condemnation of imperialism in geography is directed
at no individual; the science as a whole is to blame.

J. Blaut (1969)

▶

DOI: 10.1057/9781137301758

Contents

List of Illustrations

Figures

Tables

DOI: 10.1057/9781137301758

Preface

What does it mean to think geographically? What constitutes geographical thought? What defines geography as a discipline?

Geographers do not agree on the answers to these three questions. Not at all. As quickly as one begins to answer by talking about spatiality, another speaks of scientific study of the environment, or relationships between humans and the Earth, or places, or regions, and so on. All that disagreement is not a bad thing. On the contrary. Since it is my conviction that the discipline is amidst a *polemos* that renews the urgency of these fundamental questions, it would be contradictory for me to repress the debate on the essence of geographical thought. Still, this discord makes it difficult, perhaps impossible, to answer these three elementary questions. This does not imply that any and all answers to these questions are equally valid. Rather, it means that the stakes of these questions are raised; so too the responsibility to discern stronger answers.

Consider the question, "What constitutes geographical thought?" A strong answer would, I claim, both reflect and realize a *worldliness* that is lived or practiced by the thinker responding to this question. This is because geographical thought results from reflecting upon the world and its representations, as the discipline's name suggests. Geographical thought always emerges out of the condition of being in the world; it is neither disembodied nor ahistorical, and it can never be reduced to tables of data. It exists only because of a given thinker's engagements with the world. Because strong thought is coherent, capable of integrating distinct

DOI: 10.1057/9781137301758

positions, and born out of the critique of earlier thought,[1] it follows that strong geographical thought cannot be disentangled from the struggles inherent in worldliness, i.e., necessarily part of being in the world. Hence, geographical thought derives from worldly *polemos*.[2] Thus conceived, geography qua discipline is not the same as geographical thought. The shifting boundaries that surround the discipline called "geography" are the effects of geographical thought having being *disciplined*, i.e., regularized, institutionalized, enabled, and constrained.[3]

Now let us consider another, more conventional, answer to the questions with which I opened this chapter. It comes from Isaiah Bowman (1878–1950), more specifically his *Geography in relation to the social sciences* (1930), a work intended to put such questions to rest and secure geography's place in US higher education.[4] For Bowman, geography consists of exploring, locating, and measuring (see epigram). Bowman contends that geography—in a trope that has been repeated by many geographers—is something which is *done*, i.e., a *practice*. How is one to "do" geography? Bowman says: get out there and find something or some place; then study it, capture it, map it. By this conception, geographers strive for an objective view of the world (or at least some place or region); answers, not questions. It follows that many people are geographers these days, insofar as we carry devices—GPS units, computers, and cell phones—that can locate us spatially with extraordinary precision. With such devices ready at hand, we need only an explorer's spirit to do geography.

These are only sketches of two possible answers to the question of geography and there are surely others. Yet at the risk of seeming unduly polemical,[5] I assert that these two positions stand opposed today and their mutual negation threatens to burst our discipline asunder. Either geography is something that emerges out of confronting being in the world, i.e., through critically encountering worldliness, or it is something that takes the world for granted and proceeds by exploring, measuring, and mapping it. The former position, which we might call "critical," "hermeneutical," or "ontological," has been advanced in many forms; I will fend for it in the name of a *postcolonial* critique. The latter position, represented by Bowman, could be called *empiricism*.[6] It is the object of my critique.

Notes

1   On the relative coherence of a conception of the world, see Gramsci (Q11§12 1971 pp. 323–343); Wainwright (2011).

DOI: 10.1057/9781137301758

2   This implies that Marx and Engels' famous claim in the opening of the
     Manifesto (1848)—"The history of all hitherto existing society is the history
     of class struggles"—is also valid for the geography of our world (presuming,
     i.e., that we interpret "class struggles" in an expansive sense). This work is
     therefore motivated, in part, as a Marxist effort at elucidating a history of
     struggles that are shaping the world. As the poet Adrienne Rich, who died
     shortly before I completed this book, writes in one of her wonderful essays:

     Sometime around 1980 I felt impelled to go back and read what I had dis-
     missed or felt threatened by: I had to find out what Marx, along the way of his
     own development, had actually written. I began working my way through
     those writings, in the assorted translations and editions available to me, an
     autodidact and an outsider, not an academic or post-Marx Marxist. There
     were passages that whetted my hunger; others I traversed laboriously and in
     intellectual fatigue. [...] What kept me going was the sense of being in the
     company of a great geographer of the human condition, and specifically, a
     sense of recognition: how profit-driven economic relations filter into zones
     of thought and feeling. Marx's depiction of early 19th-century capitalism
     and its dehumanizing effect on the social landscape rang truer than ever
     at the century's end. Along with that flare of recognition came profound
     respect and empathy for Marx's restless vision of human capacities and the
     nature of their frustration. I found no blueprint for a future utopia but a
     skilled diagnosis of skewed and disfigured human relationships. (Rich
     2001, p. 38)

     I share Rich's sentiment that we should keep returning to Marx so that we
     remain in the company of "a great geographer of the human condition."
     Part of what makes Marx—and Adrienne Rich—so important for us is their
     capacious conception of the world and its "disfigured human relationships";
     so too their ability to describe the world in "poetic language untethered from
     the compromised language" of the state, media, and capital (p. 40).
       My conception of geographical thought is also inspired by Heidegger.
     Consider his essay, "The age of the world picture" (1938), which argues
     that the essence of the modern age lies in the production of the world as
     representation. Most of the essay is concerned with examining the place
     of science, as research, in producing this aspect of modernity. What is
     the essence of science as research? Heidegger writes: "Every science is, as
     research, grounded upon the projection of a circumscribed object-sphere"
     (p. 123). Such projection is invariably particularized "into specific fields of
     investigation," such as disciplines; this is not "a necessary evil, but is rather an
     essential necessity of science as research," which is essentially "*Betrieb*"
     (p. 124), meaning "industry, activity" but also "management." What links
     this peculiar drive of modern science with the problematic of representation
     is that "Knowing, as research, calls whatever is to account with regard to

DOI: 10.1057/9781137301758

the way in which and the extent to which it lets itself be put at the disposal of representation" (p. 126). The surging activity of science as research is not without consequences, since "Nature, in being calculated in advance or to verify a calculation, and history, in being historiographically verified as past, become, as it were, 'set in place' [*gestellt*]. Nature and history become the objects of a representing that explains" (pp. 126–127). Militant empiricism is a mode of setting-into-place.

3   On the disciplinarity of the social sciences, see Foucault (1966), Wallerstein et al. (1996), and Ismail (2005).

4   This dual task has generated many of the landmark works in twentieth-century Anglophone geography. Among the best-known are Hartshorne (1939), Harvey (1969), and Gregory (1978). On Bowman's study see Smith (2003, pp. 220–222). I agree with Smith's assessment that *Geography and the social sciences* is a "strangely cavalier and insecure book" (2003, p. 221).

5   This book is inherently polemical because it reflects thoughts on war composed during wartime. "Polemic" comes from a Greek word for "war" or "confrontation" which is usually transliterated *polemos*. The critical purchase of the claim that this (or indeed any) text is "merely polemical" derives from a distinction between war and not-war that has been indefinitely suspended (Agamben 2005). Moreover, *polemos* is unavoidable as a mode of being, or better, a name for Being. I take this phrase from the introduction to Fried's book on Heraclitus, Heidegger, and *polemos*:

> Being is polemical, but not in the conventional, petty sense of the term, in which a "polemic" means a refusal to take the opponent seriously in a fundamental challenge to our interpretation of the matter at hand. In reviewing what follows, the reader may well ask: But what is not polemos, given the "ontological" breadth of this account? The brief answer is that Heidegger's polemos has a scope as broad and as deep as his whole think-ing, for it describes not only our own Being, what he calls Dasein, but also Being itself. Polemos is a name for Being. (2000, p. 16)

6   Empiricism holds a complex place in the history of Western philosophy (for a survey, see Woolhouse 1990). While it is beyond the purview of this book to carefully map this history, a word may be useful on the meanings I aim to suggest by using the term here. In the first instance, empiricism is a school associated with a quartet of Anglophone thinkers—Bacon, Hobbes, Locke, and Hume—whose texts unite around the proposition that the basis of truth lies in *experience*, deriving more narrowly from data qua *sense-experience*, particularly as it is organized through *experimentation*. This school is often seen retrospectively as a middle point between skepticism (after Descartes) and Kant's synthesis; but insofar as sense-experience and the gathering of data remains the byword of scholarship in the social sciences, geographers are not so far from Locke and Hume. The critique of empiricism derives from Kant's

DOI: 10.1057/9781137301758

Critique of pure reason, which begins with the claim (1787, §I.1) that "though all our knowledge begins with experience, it by no means follows that all arises out of experience. For, on the contrary, it is quite possible that our empirical knowledge is a compound of that which we receive through impressions, and that which the faculty of cognition supplies." Subsequent criticisms of empiricism—from Hegel to Marx, Heidegger to Derrida—have extended Kant by questioning the way that empiricism thematizes the constitution of the true such that sense-experience is privileged (as unmediated, extra-ideological, merely ontic, pre-textual, and so on). Gregory's critique crystallizes these matters for geographers (1978, pp. 54–55). Nevertheless, empiricism persists. Consider the present-day devotion in geography to *experience, fieldwork*, and *data* (ocular- or lens-sensed, transcribed via computer). These tend to abut the notion that geography is essentially a practice. In other social sciences—particularly economics and political science—empiricism braces the common sense rational choice episteme; less so in geography today. At a general level the idea of "social science" is implicated in empiricism, and vice versa. Derrida writes: "empiricism is the matrix of all the faults menacing a discourse [i.e., social science,...] which continues [...] to elect to be scientific" (1966, p. 288).

A socio-historical remark may be warranted here. Bacon, Hobbes, Locke, and Hume were deeply concerned with a trinity of theoretical problems: sovereignty, property, and social progress; indeed, to generalize, they are best known for this work, not empiricism per se. I hypothesize that their conception of sense-experience, truth, reason, sovereignty, property, and social progress are tightly interlinked in and through empiricism and, for that matter, the British empire. While I cannot develop the argument here, this book is partly motivated by the intuition that there is a lingering relationship between empiricism and empire. I speculate that the becoming-empirical/imperial of the world—a process described by Heidegger's (1938) analysis of *Gestell*—is not only rooted in modernity's grasping of truth through research after Descartes' turning of Christian metaphysics (as Heidegger explains), but also a result of the calculation of the world for empire in a fashion that draws upon the empiricists' privileged synthesis of sense-experience, sovereignty-property, and calculative reasoning.

Regardless of the merits of these speculations, I trust that my reader will recognize that a critique of empiricism does not mean the mere *rejection*—if such a thing were possible—of what are called today "empirical facts." This book contains such things. The question is how we conceptualize them, live with them.

DOI: 10.1057/9781137301758

Acknowledgments

The first draft of this manuscript was written in Marcie Jacobson's *cuarto obscuro*. I thank her and Helga for this space, Diane and Kristin for the time.

The department of geography at the University of Minnesota graciously hosted a workshop to discuss a draft of this manuscript; my graduate seminar at Ohio State University read another. I thank these colleagues and critics for their insights.

In addition, for their support and criticism, I thank Kiran Asher, Josh Barkan, Niels Barmeyer, Geoff Boyce, Joe Bryan, Mat Coleman, Raymond Craib, Kiado Cruz, Oliver Fröhling, Marcus Green, Qadri Ismail, Vidhya Jayaprakash, Hyeseon Jeong, J P Jones, Will Jones, Seung-Ook Lee, Sallie Marston, Kendra McSweeney, Kristin Mercer, Tad Mutersbaugh, Kat O'Reilly, Pavel Punk, Paul Robbins, Dan Sui, Abdi Samatar, Simón Sedillo, Gayatri Chakravarty Spivak, Mary Thomas, Jacqueline Vadjunec, and my especially thoughtful anonymous reviewers.

A small grant from the Center for Latin American Studies at Ohio State University facilitated my travel to Oaxaca in July 2011. Apart from this, the research received no specific grant from any funding agency.

DOI: 10.1057/9781137301758

1
Letters from Oaxaca

Abstract: *A letter from the Sierra Juárez ~ The* México
Indígena *project ~ An accusation of "geopiracy" ~ The
Tiltepec letter ~ On the etymology of 'piracy' ~ An ancient
pirate's reply to Alexander the Great.*

Keywords: Bowman expeditions; geopiracy; *geopiratería*;
Zapotec community of Oaxaca

Wainwright, Joel. *Geopiracy: Oaxaca, Militant
Empiricism, and Geographical Thought.* New York:
Palgrave Macmillan, 2013. DOI: 10.1057/9781137301758.

On January 14, 2009, the Union of Organizations of the Sierra Juárez of Oaxaca (UNOSJO) published an open letter to criticize a research project conducted by a team of geographers under the banner of the "Bowman expeditions."[1] In their letter, UNOSJO alleges that the geography professors who ran the project—which involved mapping several rural Zapotec communities in the *Rincón de Ixtlán* of Oaxaca, Mexico—had failed to inform the communities that their research was funded by the US Army. Allow me to quote the UNOSJO letter at length:

> Towards the end of 2008, the results of the research project *México Indígena* [the name of this Bowman expedition] were handed over to two Zapotec communities in the Sierra Juárez [. . .]. Research had been undertaken two years earlier by a team of geographers from University of Kansas. What initially seemed to be a beneficial project for the communities now leaves many of the participants feeling like victims of geopiracy. [. . .]
>
> Project leader and geographer Peter Herlihy explained [that] the project [. . .] was to document the impacts of PROCEDE [a Mexican Government program that encourages the privatization of *ejido* and other community-managed lands][2] [. . .] on indigenous communities. He failed to mention, however, that this research prototype was financed by the Foreign Military Studies Office (FMSO)[3] of the United States Army and that reports on his work would be handed directly to this Office. Herlihy neglected to mention this despite being expressly asked to clarify the eventual use of the data [. . .].
>
> Herlihy mentioned that his team would collaborate with the following organizations: the American Geographical Society (AGS), [the University of] Kansas, Kansas State University, Carleton University, the *Universidad Autónoma de San Luis Potosí* and the Secretary of Environment and Natural Resources [of the Mexican state]. He failed, however, to acknowledge the participation of Radiance Technologies, a company that specializes in arms development and military intelligence. [. . .]
>
> *México Indígena* forms part of the Bowman Expeditions, a more extensive geographic research project backed and financed by the FMSO, among other institutions. The FMSO inputs information into a global database that forms an integral part of the Human Terrain System (HTS), a United States Army counterinsurgency strategy designed by FMSO and applied within indigenous communities, among others. [. . .]
>
> In November 2008, the *México Indígena* Project completed the maps corresponding to Zapotec communities San Miguel Tiltepec and San Juan Yagila. Contrary to the often-mentioned promise of transparency, *México Indígena* created an English-only web page, a language that the participating communities do not understand. Before the communities received the

DOI: 10.1057/9781137301758

work, said maps had already been published on the Internet. Furthermore, the communities were never informed that reports detailing the project would be handed over to the FMSO.

In addition to publishing the maps, the *México Indígena* team created a database into which pertinent information was entered: community member names and the associated geographic location of their plot(s) of land, formal and informal use of the land and other data that cannot be accessed via the Internet.

According to statements made by those heading the *México Indígena* research team, this type of map can be used in multiple ways. They did not specify, however, whether they would be employed for commercial, military or other purposes. [...]

UNOSJO [...] is against this kind of project being carried out in the Sierra Juárez and distances itself completely from the work compiled by the *México Indígena* research team. We call upon indigenous peoples in this country and around the world not to be fooled by these types of research projects, which usurp traditional knowledge without prior consent. Although researchers may initially claim to be conducting the projects in "good faith," said knowledge could be used against the indigenous peoples in the future.

We hereby demand that Peter Herlihy honor his promise of transparency and that the Mexican public be made aware all his sources of funding and the institutions that received information on findings obtained in the communities.

We further demand that, in light of these facts, the Mexican Government, firstly the Secretary of Environment and Natural Resources for having financed part of the research, as well as the Department of Internal Affairs, the Department of External Affairs, Deputies and Senators for possible violations of the Indigenous Peoples' National Sovereignty and Autonomy, clarify its position on the matter. (UNOSJO 2009)

Two months later, the community of San Miguel Tiltepec, Oaxaca, released another statement condemning the project. This text, the "Tiltepec letter," also deserves close reading:

The citizens of the community of San Miguel Tiltepec, through our Municipal Authority and the Authority of Communal Lands, wish to present to the public our position regarding the research project called *México Indígena*, begun in the year 2006 and ended in July 2008, which made a map that contains names of places and other cultural and geographical information provided by people from our community.

The researchers [...] presented themselves to the General Assembly of our community. They only informed us that the goal of their research was to find out the impacts of the government program *Procede* on indigenous communities. They never informed us that the data they collected in our community would be given to the Foreign Military Study Office (FMSO) of the Army of the United States, nor did they inform us that this institution was one of the sources of financing for the project. Because of this, we consider that our General Assembly was tricked by the researchers, in order to draw out the information the[y] wanted.

The community did not request the research[;] it was the researchers who convinced the community to carry it out. Thus, the research was not carried out due to the community's need, it was the researchers of the project *México Indígena* who designed the research method in order to collect the type of information that truly interested them. [...]

[W]e wish to express to the public [...] our complete disagreement with the research carried out in our community, since we were not properly informed of the true goals of the research, the use of the information obtained, and the sources of financing.

Our demand is to those responsible to the project *México Indígena*, the American [Geographical] Society, the Foreign Military Study Office of the Army of the United States, the Autonomous University of San Luis Potosí and University of Kansas, as well as all the other institutions involved, about whose participation we do not have information. We demand that:

▶ they refrain from using in any way the information they collected in our community.
▶ they return all the information they obtained from our community.
▶ they immediately destroy all the information they have on our community, and that they provide proof of having done so.
▶ they immediately eliminate from the internet all the information they published regarding the research carried out in our community.
▶ they offer us a public apology for having violated our rights as indigenous peoples, and for having violated their own norms, set out in the ethics code of the American Geographic[al] Society which they claim to respect.

Finally, we call out to the communities and indigenous peoples of Mexico and the world, for them not to be taken unawares by researchers of the Bowman expeditions, or by other researchers who only follow their interests or those of the people they represent. It is the communities and peoples themselves who should decide what they want to have researched about themselves, and who should carry it out. (Hernández and Montaño Mendoza 2009)

Within the space of three months in early 2009, geographers were thus presented with two detailed criticisms of a particular geographical research project. Clearly these letters target the "researchers of the Bowman expeditions"; yet their purview is broader, their implications profound. I think they deserve careful re-reading as commentaries on geographical research. These letters constitute extremely rare statements concerning how research should and should not be conducted *from the point of view of the research subjects.* Moreover they situate their critique historically and geographically (not to mention politically) by placing their critique of this particular project within a multi-scalar analysis of the research. This explains a duality in the tone of these texts, which move between precise statements about events that occurred in their communities and open-ended comments on things far beyond *Sierra Juárez.*[4] Consider for instance that the five demands made in the Tiltepec letter are made not only to four institutions, but also to "all the other institutions involved, about whose participation we do not have information." (Who or what are these "other institutions"? Do we geographers "have information" about them? Was the Association of American Geographers (AAG), for instance, "involved"?) In a general sense the letters imply or diagnose a conflict between epistemic communities, one involving institutions, conceptions of the world, and political forces—not merely the actions of a few individuals. Emphasizing these institutional and political qualities implies a strategy of de-personalizing the critique, expanding it beyond the immediate target (the "researchers of the Bowman expeditions"). Likewise the Tiltepec letter's conclusion—a "call" to "communities [around] the world [to avoid being] taken unaware by researchers who only follow their [own] interests"—would seem to implicate, or interpellate, many geographers. It should solicit a thoughtful response.

There is much more to say about these letters, which inspired the writing of this book, but for the moment I limit myself to one final point. The final line of the first paragraph of the UNOSJO letter elegantly summarizes the critique: "What initially seemed to be a beneficial project for the communities now leaves many of the participants feeling like victims of geopiracy [*geopiratería*]." This concept is not defined, but its meaning is implied through the two texts and consistent with the neologism. Something of their world was seized, taken unjustly, plundered: the worldliness of the communities of the *Rincón*, captured by geographical pirates.

* * *

According to the *OED*, the English "piracy" (like the Spanish *piratería*) is formed etymologically from the post-classical Latin *piratia*. The noun "pirate" is of classical Latin *pīrāta*, derived from the Hellenistic Greek πειρατής, in turn from the ancient Greek πειρᾶν meaning "to attempt, attack, assault." A more ancient root πεῖρα carries the meanings "trial, attempt, endeavor"—synonyms for research today. In his erudite genealogy of the figure of the pirate, Heller-Roazen (2009) notes that the classical conception of the pirate as "the enemy of all" (the words are Cicero's) not only provides the earliest-known justification for the existence of international law, but also underlies the modern conception of the state. This is because the state (and law) exist to specify the properly piratical, i.e., to discern *which* forms of plunder are sanctioned and which are not. The notion that the state is rooted in plunder[5] is ancient. Augustine of Hippo (354–450 CE) asks:

> What are kingdoms but great robberies? And what are robberies themselves, but little kingdoms? The band itself is made up of men; it is ruled by the authority of a prince, it is knit together by the pack of the confederacy; the booty is divided by the law agreed upon. If […] this evil increases to such a degree that it holds place, fixes abodes, takes possession of cities, and subdues peoples, it assumes the more plainly the name of a kingdom.

The pirate ship bears the essential attributes of what we call the state. Augustine continues:

> [I]t was an apt and true reply which was given to Alexander the Great by a pirate who had been seized. For when the king had asked the man what he meant by keeping hostile possession of the sea, he answered with bold pride: "What thou meanest by seizing the whole earth; but because I do it with a petty ship, I am called a robber, whilst thou dost it with a great fleet art styled Emperor."[6]

With the same bold spirit of this unnamed "pirate" who confronted Alexander the Great, the indigenous communities of Oaxaca have cast the word "geopiracy" northward. And thus we are compelled to ask: Just who is sanctioned to plunder today? What justifies the power to "seize the whole earth"? How should we respond to geopiracy?[7]

Notes

1   Geographer Zoltan Grossman maintains a website with a thorough collection of links and texts on the Bowman expeditions: http://academic.evergreen.edu/g/grossmaz/bowman.html
2   The literature on PROCEDE's consequences for the privatization and consolidation of lands is considerable. For broadly supportive evaluations see

DOI: 10.1057/9781137301758

de Janvry et al. (1997) and Deininger et al. (2001); for more critical discussions see Appendini (2001), Stephen (2002), de Ita (2006), and Wilshusen (2010).

3   According to its website, the FMSO is "an open source research organization of the U.S. Army. Founded as the Soviet Army Studies Office in 1986, it was an innovative program that brought together military specialists and civilian academics to focus on military and security topics derived from unclassified, foreign media. [...] The FMSO conducts unclassified research of foreign perspectives of defense and security issues that are understudied [...] but that are important for understanding the environments in which the U.S. military operates" (accessed December 9, 2011 at http://fmso.leavenworth.army.mil/).

4   Like all communities, these are heterogeneous. I do not mean to suggest that everyone within the Zapotec community participated in writing these texts, or that all the members of the community support the arguments made in them. On the contrary, my intuition is that they were written by a relatively small group within these communities, and that there is some dissent from the views they represent. But one does not need to romantically believe in the existence of an organically unified indigenous community to read and appreciate specific claims that are made in its name by some of its leaders.

5   See Karatani (2008); Karatani and Wainwright (2012).

6   Cited in Heller-Roazen (2009, p. 56). Later in Heller-Roazen's archaeology we find that the "common enemy of all" is an exceptional opponent against whom the ordinary rules of law and warfare do not apply. But this does not mean that we should feel good about pirates. And while Heller-Roazen's analysis is extremely useful, we should not hasten to impose it upon the claim of "geopiracy" as made in Oaxaca. To clarify the relation between the "common enemy of all" and the "geopirates" would be, ipso facto, to trace the history and geography of the present: Heller-Roazen's archaeology, after all, traces the emergence of modern law to its present (exceptional) state where the US, in its effort to map the world in the name of securing all "common enemies," stimulates "geopiracy."

7   The term "geopiracy," probably first used in Oaxaca by Aldo González (then the director of UNOSJO), no doubt suggested itself because of prominent discourse on "biopiracy" in Mexico. Since the publication of UNOSJO's public letter, the term "geopiracy" has taken on another meaning, as a synonym for "geoengineering," referring negatively to the practice of manipulating the Earth's environment to counteract climate change (ETC Group 2010 and 2011). A third meaning of "geopiracy" has been proposed by Vogel et al. (2008), to refer to "the false attribution of location in the visual arts," such as when a film set in the Amazon is actually filmed in Belize. Not far behind is a television show, "Geopirates," intended to teach children about geographical diversity (www.thegeopirates.com).

DOI: 10.1057/9781137301758

2
Geographers Respond: I

Abstract: *The accused geographers reply to the critics ~ The problematic of the Bowman expeditions ~ The elusive concept of "human terrain" ~ "Let the indigenous people of Oaxaca speak for themselves" ~ Professor Herlihy's public defense ~ Isaiah Bowman redux ~ Militant empiricism defined ~ A complex interpretive challenge.*

Keywords: Bowman expeditions; human terrain; indigenous mapping; militant empiricism; military participation in academic research

Wainwright, Joel. *Geopiracy: Oaxaca, Militant Empiricism, and Geographical Thought.* New York: Palgrave Macmillan, 2013. DOI: 10.1057/9781137301758.

DOI: 10.1057/9781137301758

The accused geographers have written a number of texts that, taken together, comprise a lengthy and robust defense (cf. AGS 2009; Dobson 2006a; Herlihy et al. 2006; Herlihy et al. 2008; Dobson 2009; Herlihy 2010a, 2010b). As these dates reveal, some explanations of the Bowman expeditions—including several of the most straightforward and reveal-ing—were written before 2009, i.e., before the two letters from Oaxaca. For the geographers of the Bowman expeditions this is an important fact, since it suggests that they have been transparent about their work all along. Let us consider one illustrative text (*México Indígena* ca. 2005) which outlines the research problematic:

> *Problematic.* The Bowman Expeditions Program resolves two fundamental problems facing the development of GIS databases on different countries around the globe today:
>
> 1   Disaggregated data on the world's peoples and places are not in accurate and meaningful formats with geolocational precision for matching place with ethnicity, populations, resources, loyalties, etc.
> 2   Foreign digital geographies or "human terrains" demand accurate, on-the-ground field knowledge of the "cultural terrain."

In the archive of texts on the Bowman expeditions, this represents the clearest (and perhaps the earliest) public statement of the underlying motivation for this research. The geographers' use of the Althusserian term "problematic" seems appropriate (though it is not explained in this text), since this statement defines the intellectual horizon of the work: its definitive questions, conceptual orientation, and limits.[1] Reading their problematic, we can say that the expeditions emerged as a response to problems with "the development of GIS databases": problems, that is, of location and measurement. Taken together, the two sub-points show that the project derives from the need to *represent social difference in spatial form*. The research is motivated by the lack of "data on [...] peoples and places." Why is there a lack? What is the problem with the existing data on "ethnicity, populations, resources, [and] loyalties"? The statement of the problematic says that existing data is insufficiently "accurate," or not in the right *form*; consequently we lack the necessary "geolocational pre-cision" that would allow us to match "place with ethnicity, populations, resources, [and] loyalties." This language may sound familiar to geogra-phers with experience in cultural ecology and indigenous mapping, which has long prioritized the task of mapping cultural regions (ethnicity) with traditional and customary resource uses.[2] Hence it is unsurprising that

DOI: 10.1057/9781137301758

the Bowman expeditions have drawn considerable support from those cultural ecologists with experience in participatory mapping projects and who trace their ideas to Carl Sauer—a geographer whose expeditions to Latin America were noteworthy for their exacting concern with capturing cultural differences in spatial form (as noted in Herlihy 2010a).

Let us restate the problematic of the Bowman expeditions more clearly. The mandate of the Bowman expeditions stems from the lack of spatial data—accurate, precise, properly formulated spatial data—that would allow X to know, match, and map *difference*: different people and places and the relationships between them. All this is clear, except for one thing: a mandate such as this presupposes a *subject,* the X in my restatement of their problematic. Accurate, precise, spatial data—*who* lacks this? That is the question. Their research mandate is dependent upon this subject.

The purpose of point (2) is to fulfill this mandate by defining X. Unlike point (1), point (2) is not so clear. Let us examine it more closely. The sentence's verb—to "demand"—is strong. The problem lies with the sentence's subject: "Foreign digital geographies or 'human terrains.'" The "or" implies that this is one subject, which (or who) may assume two forms. This two-headed subject would appear to be the X implied in (1), that being which "demand[s] accurate, on-the-ground field knowledge" who needs accurate spatial data on different "peoples and places." Yet what are "foreign digital geographies or 'human terrains'"? What is this two-headed subject that has resurrected Isaiah Bowman from the grave to christen their expeditions? What sovereign bid them to set sail with the flag of Bowman's tripartite formula—explore, locate, measure—set upon the mast?

Alas, the text does not say. And since the concept "human terrain" is not commonly used in the human geography literature,[3] we will have to search elsewhere for an answer. But first let's read two detailed, public replies to the two letters from Oaxaca.

* * *

The first is a forceful essay by Professor Dobson entitled, "Let the indigenous people of Oaxaca speak for themselves" (2009). The text reads like a bulleted memo, firing through eight "salient points"; for my purposes I will recapitulate only two (for the others, see Dobson 2009). First, Dobson contends that "the Bowman Expedition team [was] open and honest about who funded this research.... Every detail has been on the

DOI: 10.1057/9781137301758

México Indígena and AGS websites from the very beginning. It's been announced in every presentation and publication" (p. 2). I should note that all the texts I discuss in this section—and most of the documents that have been cited by the critics of the Bowman expeditions—were posted on the web either by the University of Kansas geographers or by the US Army-FMSO. There is more to the question of transparency than this, of course. We might ask, for instance, whether people in rural Oaxaca might reasonably be expected to examine the English-language websites of all those who visit their communities. But we cannot fault Professors Dobson or Herlihy for failing to disclose the US Army funding of their research on the web.[4]

The second important element of this text concerns Dobson's narrative of the origins of the Bowman expeditions. Allow me to quote his explanation at length:

> My whole rationale for Bowman Expeditions is based on my firm belief that geographic ignorance is the principal cause of the blunders that have characterized American foreign policy since the end of World War II. [...] I was incredulous when I first heard rumblings that the Bush Administration was thinking of attacking Iraq. During the debate, such as it was, I never thought the case for weapons of mass destruction was convincing whatsoever. [...] I believed the misinterpretation was due to geographic ignorance [...]. I felt it pervaded, not just the White House, but Congress, analysts [...], journalists, and the public as well. There is plenty of evidence that President Bush himself did not understand what he was getting us into. Ambassador Peter Galbraith reported that Bush didn't even know there were two sects of Islam until shortly before the invasion began. That is geographic ignorance at the highest level. [...]
>
> Convinced that geographic ignorance has been the cornerstone of U.S. foreign policy since the end of World War II, I asked myself what the American Geographical Society could do about it. I conceived of sending a team of geographers to every country in the world to improve geographic understanding, connect with scholars, and bring back that knowledge to the American people. I did a calculation and was astounded to realize that it would cost only $125,000,000 to send a professor and two or three graduate students to every country in the world to spend a full semester every year. [...] So far, we've received about $2,500,000, a good "down payment," but far less than what's needed to make a sizable dent in the America[n] scourge of geographic ignorance. This is the noble effort that [UNOSJO's then-director, Aldo] Gonzalez is trying to quash in one part of the world, in direct opposition to the people who live there.

DOI: 10.1057/9781137301758

We could quibble with the lack of evidence offered that Aldo González acted in "direct opposition" to the people of the region.[5] But let us not get lost in this particular ad hominem. At the risk of oversimplification, allow me to recapitulate the broader narrative: the US invasion of Iraq was an error, grounded in a lack of geographical knowledge by elites— not only in the state (White House, Congress) but also civil society (journalists, analysts). Ergo Dobson decided to use his position as the president of the AGS to foment a campaign to train more geographers who could produce "useful information" and, fortuitously, the US government responded favorably.[6] Critics have tried to stop this 'noble effort' (no explanation of their motivations is offered).[7] The suggestion is that González and other unnamed critics are irrationally holding back the progress of science in overcoming geographic ignorance. The result of such ignorance is the continued failure of US foreign policy. It is strongly implied that the critics are hurting the United States.

Most aspects of this narrative are repeated by Professor Herlihy in his writings on the Bowman expeditions (cf. 2010a, 2010b). Yet, there are some noteworthy differences in emphasis in the intertwining conceptions of their work. For instance, Dobson's texts tend to characterize the Bowman expeditions as, on one hand, good, old-fashioned, empirical, fieldwork-based geographical research, and, on the other, a crucial tool for the US state and its military. In Herlihy's writings this pair is complemented, if not overshadowed, by a third motif, i.e., that the Bowman expeditions comprise a progressive political intervention on behalf of Latin America's indigenous peoples. In what is perhaps best characterized as a liberal-interventionist framing (not to mention a colonial trope), the Bowman expeditions are explained as a project to help indigenous people. Consider Herlihy's 2010 public presentation at the annual meeting of the AAG—a vociferous defense of his work:

> Shaped by ethical decisions concerning what information local people do or do not want to collect and share with outsiders, [the participatory mapping projects that Herlihy has worked on, and which he reviewed in painstaking detail during his presentation], including *México Indígena*, were designed from a deep ethical commitment to help indigenous people, not hurt them. Our focus is the legitimization of popular knowledge and its conversion to scientific knowledge, *with the objective of contributing to a science of the proletariat* through which the masses are able to conduct their own fights with their own maps for social transformation and justice. And for our armchair critics waxing on the dangers of military funding to indigenous peoples,

DOI: 10.1057/9781137301758

each of the [mapping projects discussed by Herlihy] has indeed *included military participation.* (Herlihy 2010a, my italics)

Arriving as they did at the end of a long and otherwise boring presentation, these concluding remarks electrified the audience (I was present).[8] They interlace discursive elements that do not typically coincide: a celebration of *indigenous mapping* as an anti-colonial practice[9]; a defensive posturing in favor of *military participation* in research; and an appeal to transform popular knowledge into a *revolutionary science.* Those familiar with the Marxist tradition will recognize this final theme, which bears a controversial legacy. Paired as it is here with a fierce pride in military collaboration, these lines cannot but remind us of Josef Stalin's fateful fusion of populism, militarism, and the cult of science (see Deutscher 1949).

For a more precise reading of these texts it is helpful to examine them alongside Neil Smith's patient biography of Isaiah Bowman, an "academic entrepreneur" (2003, p. xx) who sought to put geography to work in building the American empire. Smith's study describes Bowman, in distinct but overlapping terms, as a "liberal internationalist" (p. 26); a "liberal pragmatis[t]" (p. 75); and a "pragmatic positivis[t]" (p. 185). Put these together and you get "liberal internationalist pragmatist positivist," which nicely captures the ethos of the Bowman expeditions. As we will see, the texts of the Bowman expeditions faithfully reiterate Bowman's ideology. To put this otherwise: the implicit theory of the expeditions conforms to the name given to them.[10]

Beyond this thematic continuity, the Bowman expedition texts make two core claims that we may treat as fundamental to their problematic.[11] The first of these assertions is that the epistemic mandate of the Bowman expeditions stems from conditions inherent to the United States qua nation-state that require a renewed investment in the discipline. These conditions, which are not so much analyzed as asserted, are principally characterized as political and military in character. A corollary of this axiom is that the expeditions should help the US state/military (as well as the communities under study: but the latter point is not uniformly emphasized, so I do not treat it as axiomatic). To briefly cite two illustrations: remember Dobson's claim that the origin of the Bowman expedition lay in his conviction "that geographic ignorance has been the cornerstone of US foreign policy since the end of World War II, I [...] conceived of sending a team of geographers to every country in the world to improve geographic understanding, connect with scholars, and bring back that knowledge

DOI: 10.1057/9781137301758

to the American people." (In this discourse we may see the distinction between the needs of "US foreign policy" and the geographical knowledge of "the American people" is blurred to a point where it becomes opaque.) A second illustration is provided by Herlihy et al.'s remarkable argument that "foreign intelligence is geography, and *geographers stand ready to assist the nation* today as their predecessors did so effectively in the past" (2006, p. 4, my italics). This statement could be interpreted as a performative claim or a presumptuous and nationalist one. In either case we should ask how it is that these authors feel they can speak authoritatively for all geographers—and which precise "nation" they wish to assist.

The second claim that I take as axiomatic—made principally, and repeatedly, by Dobson (e.g., 2005a, 2006c, 2010)—is that investment in geographical research today should prioritize foreign, field-based research of "human terrain" and its integration via GIS. This point could be construed as methodological, but as I argue, the essence of this claim goes deeper than mere method. This is because (to anticipate my argument) their very *conception* of the need for "geolocational precision for matching place with ethnicity" and for "digital geographies or 'human terrains'" linked with "*accurate, on-the-ground field knowledge*" is grounded upon presuppositions that are both strategic (in the military sense) and ontological (i.e., concerning the character of being in the world).

Since what follows can only be construed as a critique of their claims, let me say this. I do not wish to underestimate the logic behind the Bowman expeditions. Some geographers may disagree with these two premises, as I do, but it would be unwise to ignore their coherence and, for many, their persuasiveness. In what follows I do not so much question the inner solidity of their premises, but rather inquire into the binding rigor—that which secures their coherence. Let me go further. We should grant *in a certain way* the argument that the sort of "accurate, on-the-ground field knowledge" that empirical geographical research provides *is* in fact urgently needed by the US state/military today. Our ethical conundrum concerns *how we address this need*. I call the enthusiastic embrace of this need *militant empiricism*.

* * *

Let me briefly recapitulate. We have, at least superficially, a two-sided tale from the field: the *researchers* say that their subjects (and Institutional

DOI: 10.1057/9781137301758

Review Board (IRB) or anyone with internet access for that matter) could have known their problematic, sources of funding, and research program. Their research subjects—the *researched* of Oaxaca—contest this point fundamentally: they did *not* know, they claim, that the geographers were funded by and collaborating with the US Army; nor did they concede to share the maps of and data about their communities with the US Army. Something of their world was plundered. They are victims of geopiracy.

This dichotomy presents us with a complex interpretive challenge. How should we evaluate these opposing positions? How should geographers who were not involved in the research (and perhaps who have never been to Oaxaca) respond to the debate? Should we evaluate these claims at all? With which warrant? Is such an evaluation a problem for geography qua discipline? For geographical thought? How might such an evaluation proceed? Would such an evaluation itself constitute geographical research? Would it contribute to geographic knowledge? (Perhaps even enough to merit publication in the *Annals*?) Is there, indeed, *anything* particularly "geographical" about this dispute at all? Or is it merely coincidental that this conflict is playing out in our discipline? (And perhaps, in this event, we might justly hope it will all go away?)

I claim that *these questions merely restate those with which I opened this study.* They only reformulate them in a more recognizably political purview. For this reason the stakes of the basic questions—what is geography?, etc.—are clarified and intensified.

To shed further light on these questions, let us now consider how geographers in the United States responded to the situation when, in early 2009, we were presented with these competing claims about the Bowman expedition to Oaxaca.

Notes

1   On Althusser's use of problematic (*problématique*) see *For Marx* (1965), where Brewster's glossary defines problematic as the "theoretical or ideological framework" in which a concept exists and is used meaningfully (pp. 253–254). Brewster stresses that we consider "the *absence* of problems and concepts within the problematic as much as their presence" (my italics).

2   The literature on the politics of cartography has grown too vast to review here. On the politics of mapping indigenous Mexico, see especially Craib (2004); on the politics of recent indigenous counter-mapping projects in Latin America, compare Nietschmann (1995); Herlihy and Knapp (2003);

DOI: 10.1057/9781137301758

Chapin et al. (2005); Bryan (2007); Wainwright (2008, chapter 6); and Wainwright and Bryan (2009a).

3 A search of "human terrain" and "cultural terrain" in the recent literature on human geography found no references (apart from citations of uses by the US military).

4 Likewise I think we can take their word that the IRB at the University of Kansas was aware of the source of their funds and did not hold up their research on these grounds (Dobson 2009; Herlihy 2010a). As IRBs exist principally to legally protect US universities, I see no reason to expect them to check militant empiricism. On IRBs and social research see AAUP (2000). On the politics of intellectual property rights and indigenous research see Madson (2008).

5 There is more to be said on the involvement of Aldo González, UNOSJO's former director, in Herlihy's initial research steps in Oaxaca, but I will withhold comment because (although I have heard the story from both sides) neither party has published a text (to my knowledge) explaining this relationship. The key point of Dobson's critique is that González unethically inserted himself between the geographers and the community, effectively claiming to represent the indigenous peoples *himself*. In his essay Dobson calls out several would-be "representatives"—UNOSJO, González, and unnamed "armchair critics" (including myself no doubt)—to insist that the indigenous people "speak for themselves." Yet we must note that Dobson's text is written precisely in order to speak for the indigenous people of Oaxaca. Moreover, González is himself an indigenous Oaxaceño, so when he speaks he could be seen as speaking *from* an indigenous Oaxacan community, and (based upon our engagements) he seems exceptionally careful about not claiming to speak for "the indigenous people." In other words, the title of Dobson's essay alone seems to be sufficient to deflate his critique of González.

González no longer directs UNOSJO. After the 2010 elections in Oaxaca ejected the PRI from power, he took up a position with the state government in the Ministry of Indigenous Affairs. Hence we could say that, today, González officially "represents" the indigenous people of Oaxaca in the state, which raises the question: When González condemns the Bowman expeditions today, is his critique more legitimate for the fact that he occupies this subject-position?

6 According to Professor Herlihy's 2008 curriculum vitae, in 2005–2007 he and Professor Dobson received $1.07 million from the US DoD, State Department and the Mexican state (~60% of this sum came from the US Army's FMSO). Since 2008 they have received more funds from FMSO for expeditions to Colombia and Jordan, plus research in Honduras and follow-up work in Kansas. Given Dobson's (2009) statement, it seems that

DOI: 10.1057/9781137301758

Professors Dobson and Herlihy received US$2.5 million between 2005 and 2009. If these figures are accurate, between 2005 and 2007 they received ~$1.07 million, and in 2008–2009, another ~$1.43 million. These are not insignificant figures for our discipline. Yet we should be careful not to overemphasize money when we try to explain the motivations of those involved; ideology, disciplinary power, and even status anxiety probably play a greater role.

7   This is true of all the texts cited in chapter 3.

8   Each of the panelists were asked to speak for 10 minutes, but Herlihy spoke for more than 34. A trivial point, perhaps; yet as a consequence the other speakers on the panel (including two would-be critics) had to cut their remarks short.

9   On indigenous mapping, see note 2.

10   See Smith (2003), *passim*. The prologue of *American empire* concludes by noting that Isaiah Bowman has fallen into almost complete obscurity, even within geography (p. xxii); yet, ironically, as *American empire* was in press, geographers were busily exhuming Bowman—to set him back to work for America's empire. This makes Smith's closing remark in the prologue all the more prescient: "[Bowman] is comparatively invisible today precisely *because* of the sharpness with which he expressed the contradictions of liberalism from inside [...] the vortex of power" (p. xxii, italics in original). Bowman's return exposes these contradictions anew—and makes Smith's study crucial.

11   These generalizations are not intended to foreclose a closer reading of these texts. It would be useful to return to them to produce a more patient reading than I can offer here.

DOI: 10.1057/9781137301758

3
Geographers Respond: II

Abstract: *The author's position defined via Said's "committed amateur" and Gramsci's "traditional intellectual" ~ Three themes which structure the discourse on the Oaxaca controversy ~ Military participation as geography's unknown known ~ The AAG changes its ethics statement ~ The valences of President Agnew ~ The "geographic establishment" ~ A contrast with anthropology's response to the Human Terrain System.*

Keywords: Antonio Gramsci; John Agnew; anthropology; consenting subjects; geodata; geography; military participation in academic research; "Oaxaca controversy"; polemos

Wainwright, Joel. *Geopiracy: Oaxaca, Militant Empiricism, and Geographical Thought.* New York: Palgrave Macmillan, 2013. DOI: 10.1057/9781137301758.

DOI: 10.1057/9781137301758

To this point we have considered specific texts. Here we must venture away from this path onto the terrain of generalization because, although many geographers have discussed these matters, very few have done so in writing.[1] Hence I must summarize positions staked by diverse geographers, often informally. As this text constitutes another in this line of responses, it is perhaps the place to specify the basis of my understanding and (insofar as I am aware of them) the desires that motivate my writing.

In January 2009, when UNOSJO's letter was released, I was in Mexico, where I became involved in conversations on the Bowman expeditions that continue to the present (October 2012). During this time I have discussed these matters with many of those cited in the bibliography. I have also written a pair of coauthored public statements (discussed below) and lectured on these issues. I have traveled to Oaxaca on four occasions since 2004 where, among other things, I participated in the July 24, 2011, meeting on geopiracy hosted by the Municipal Authority of San Juan Yagila (discussed below). Yet I emphatically do not consider myself an "expert" on Oaxaca, Isaiah Bowman, or the US military. I write as a committed amateur (Said 1994, pp. 73–83).[2] Edward Said's critique of expertise, and his controversial proposal that we enact and defend the critique of empire in the name of a committed amateurism, drew inspiration from Antonio Gramsci's prison notes on the political functions of intellectuals (1971, pp. 5–13). Gramsci contends that everyone is an intellectual insofar as they think, use language creatively, have a conception of the world, and so on. This does not mean, of course, that everyone makes a living as an intellectual (as traditionally defined). As Gramsci explains:

> All men are intellectuals, one could therefore say: but not all men have in society the function of intellectuals. When one distinguishes between intellectuals and non-intellectuals, *one is referring in reality only to the immediate social function of the professional category of the intellectuals*, that is, one has in mind the direction in which their specific professional activity is weighted, whether towards intellectual elaboration or towards muscular-nervous effort. This means that, although one can speak of intellectuals, one cannot speak of non-intellectuals, because non-intellectuals do not exist. [...] There is no human activity from which every form of intellectual participation can be excluded: *homo faber* cannot be separated from *homo sapiens*. (Q12§3 1971, p. 9, my italics)

What distinguishes an ordinary person qua intellectual from a "traditional intellectual," then, is one's social function. When I claim to write

DOI: 10.1057/9781137301758

as an amateur, this is not to deny that I am a "traditional intellectual" in Gramsci's sense.[3] Those of us who serve this function today tend (as I do) to be members of the proletariat, a class defined by selling our labor power to earn a wage. Within this class, traditional intellectuals often enjoy certain privileges, such as the capacity to write about controversial topics with relative liberty. With privilege comes responsibility. The thorny questions concern how we are to make good on our responsibility. As Gramsci elaborates in Q12§3 (Ibid.):

> Each man [...] carries on some form of intellectual activity, that is, he is a "philosopher," an artist, a man of taste, he participates in a particular conception of the world, has a conscious line of moral conduct, and therefore contributes to sustain a conception of the world or to modify it, that is, to bring into being new modes of thought.

Hence, this text is the work of an amateur and a traditional intellectual, one who enjoys the opportunity to read these texts and discuss them, with the aim of modifying a specific conception of the world and thereby contributing to the existence of "new modes of thought." It is worth remarking that my privileges—fluency in English, access to publishers, tenure, and so on—explain in part why this text, and not one from Oaxaca, is in your hands.

§3.1.   Three prominent themes structure the discourse on the so-called Oaxaca controversy.[4] Though intertwining in colloquial professional chatter, they reflect distinct ways of framing the controversy. Let us examine each in turn, moving from the most to the least prominent.

The first and most prominent frame treats the controversy as a dispute about *indigenous peoples* as *consenting subjects*. This theme interprets the Oaxaca controversy against the extensive (and often unjust) history of geographical research in indigenous communities in Latin America. The Oaxaca controversy, it is said here, reflects another case where geographers may not have handled the relations with indigenous people with sufficient care and thereby entered fraught ethical thickets (so easily identified retrospectively). Consider for instance three texts written by the leaders of the Indigenous Peoples Specialty Group of the AAG, or IPSG (2009a, 2009b, 2010). These texts criticize the Bowman expeditions for exploiting the indigenous communities of Oaxaca, encouraging us to respond by creating "a system of accountability for geographic researchers working with Indigenous nations and communities" (2009a, p. 1). Developing this theme further, IPSG (2010) provides suggestions and

DOI: 10.1057/9781137301758

specific questions to help geographers approach indigenous communities respectfully. The underlying principle is captured in a letter written by the IPSG to San Juan Tiltepec, Oaxaca: "researchers should be in service to you, the Indigenous community" (2009b, p. 1). Curiously, as we have seen, several of the texts by the Bowman expedition geographers state that the purpose of their research was to help indigenous communities. There are other parallels. Texts on each side of the controversy characterize indigenous peoples as essentially distinctive, inferring that geographers should approach them with special care. In the discussions around the Bowman expeditions, this trope tends to gravitate toward discussions on the proper mode of oversight and regulation of research with indigenous subjects. Hence, for instance, the IPSG implies that the appropriate means to secure a "system of accountability" is for "the AAG to begin a dialogue with the Institutional Review Board (IRB) to more thoroughly address research with Indigenous communities" (IPSG 2009b, p. 2). Likewise the texts of the Bowman expedition geographers frequently emphasize their close work with the IRB at the University of Kansas.[5] When the same geographers organized the first "World Human Geography Conference" in 2011, they selected the theme "communities and ethics," focusing on indigenous peoples. (Funding for the conference, held at Haskell Indian Nations University in Lawrence, Kansas, was provided by the US Army Research Office.)[6]

The second theme in this discourse is that the Oaxaca controversy reveals the peculiar ethical challenges of *collecting geodata* and *mapping*. The key claim here is that there is something distinctive about the collection and representation of spatial data that requires special ethical consideration (Agnew 2009, 2010; see also Crampton 2003, 2007, 2008). Consider for instance Tad Mutersbaugh's (2009) proposal for "strengthening of geodata research protocols," probably the first widely discussed text written by an Anglophone geographer after the publication of the initial letters from Oaxaca:

> [T]he *México Indígena* project [...] should serve as a wake-up call for all of us who gather geodata regardless of source of financing or disciplinary background. [...] I would advocate for:
>
> 1   A strengthening of the AAG ethics statement to specify "full disclosure" in greater detail; and
> 2   A statement to the effect that geodata remains the intellectual property of the originating peoples—*indigenous or no[t]*—and should in no case be removed from their physical control. (2009, p. 1, my italics)

DOI: 10.1057/9781137301758

> I am not interested in a censure of México Indígena or participating schol-
> ars and I would invite them to contribute to a strong geodata standard.
> However, after reading recent statements by México Indígena researchers,
> I remain unconvinced that they had a coherent and systematic plan for dis-
> closing funding sources to each and every research participant that would
> meet the spirit of informed consent.[7]

I emphasize the phrase "indigenous or no[t]" to underscore the distinction between the first two themes in this discourse.

Although Mutersbaugh's proposal failed to generate an effective response
(so far as I am aware), it captures the ethos of the second theme of the
discourse surrounding the controversy as a tale revealing the risks inherent with collecting geospatial data. In this view, the letters from Oaxaca
reveal the complexities of geographical research at a time when seemingly
anything can be accurately mapped with fast and inexpensive techniques.
In this era of "open-source" cartography—or to use Dobson and Fisher's
(2003) felicitous expression, of potential "geoslavery"[8]—geographers
have greater powers and responsibilities. As with Mutersbaugh's proposal
(2009, p. 5), these conversations tend to gravitate toward treating geodata
as intellectual property and thus to create legal protocols for its protec-
tion (see Sui 2007). Just as the emphasis on indigenous subjects implies
the need for regulation by IRB, the emphasis on geodata suggests the need
for the regulation of geographical research by property law. I emphasize
that, again, key notions are shared on both sides of the controversy (e.g.,
Dobson 2003).

The third theme centers upon the question of *military involvement in,
and funding of, geographical research.* The claim here is that the letters from
Oaxaca reveal that military involvement (in this case by the US Army)
presents unique complications for geographical research. Consider for
instance two texts published by Joe Bryan. A letter addressed to AAG
President John Agnew and signed by 26 geographers (including myself)[9]
elaborates upon the role of the military in the design and character of
the projects:

> According to [Bowman expedition] project reports, representatives from
> the [US Army's Foreign Military Studies Office] played an important role
> in the design and implementation of their research. [...] Herlihy and his
> colleagues performed much of their research in Oaxaca under contract
> with the FMSO. The contract was managed by Radiance Technologies, an
> Alabama-based military contractor specializing in preparing intelligence
> and operational logistics.[10] [...] Other documents explicitly indicated

DOI: 10.1057/9781137301758

that Herlihy's research is relevant to FMSO and, by extension, US Army Intelligence. For instance, the July 2005 progress report filed by Herlihy for the *México Indígena* project explains that they are "constructing a very broad national-level GIS that the FMSO would find useful in many different types of analyzes [sic]"[...]. Other reports to FMSO outline potential applications of the *México Indígena* project methods for Iraq. These and other project documents warrant further scrutiny of Herlihy's claim that the *México Indígena* project was "not designed in any way for military applications." [...] Even if the *México Indígena* project is as transparent and scholarly as Herlihy claims, he fails to recognize that even the perception of impropriety constitutes a major setback for all who do similar kinds of research[...]. We therefore urge the AAG to conduct an inquiry into the *México Indígena* project [...and] investigate (1) the evidence that Herlihy revealed his funding source at the time of obtaining consent; (2) the extent that the FMSO shaped the design of the research itself; and (3) the extent to which Herlihy has made the results from the research available to FMSO personnel. (Bryan and Wainwright 2009, p. 2)

In a subsequent essay, Bryan (2010) elaborates on some of these arguments and draws this definitive conclusion:

> It is one thing for the military to search scholarly sources for intelligence. It is quite another to gather intelligence for the military on [...] indigenous peoples—a practice explicitly banned by the AAG's Statement of Ethics (AAG 2009). The *México Indígena* project appears to have done [the latter], taking advantage of indigenous peoples' desires for land rights to gather intelligence that will let policymakers more effectively intervene in indigenous affairs. In doing so, it appears to further violate provisions in the AAG's Statement of Ethics stipulating that "the dignity, safety, and well-being of informants and local colleagues" should take precedence over research goals (AAG 2009). [And yet, t]he AAG has rejected calls for an investigation, claiming that it lacks the standing to investigate its members [...].[11]

Bryan's texts emphasize the military aspect of the controversy in a fashion that suggests that military funding cannot be separated from the previous two elements. Indeed, the central thesis of his *Political Geography* essay is that the controversy "goes beyond allegations of misconduct and failure to obtain informed consent. It also raises broader questions about the relationship between academic research, military intelligence, and power relations" (2010, p. 1). The implication is clear. The controversy *concerns* indigenous people, informed consent, the collection of geodata, and so on—but these are only proximate issues. The underlying *cause*

DOI: 10.1057/9781137301758

of the conflict, and what makes these elements problematic, is the US military's involvement and funding.

Unlike the first two elements of this discourse, there seem to be no common points of agreement on the two sides of the controversy concerning this theme—only open *polemos*. I cite Bryan's texts here because they are the only published texts by an Anglophone geographer that consistently emphasize this third element of this discourse. This is not to suggest, however, that geographers have collectively missed this third aspect of the controversy. On the contrary. This third element has thematically organized the professional chatter surrounding the controversy—yet it remains the repressed center of the discourse. The military's participation in funding and shaping research is, to cite Žižek's wicked reading of Rumsfeld's logical categories (2004), our disciplinary *unknown known*.

§3.2. The institutional mechanisms of the AAG became drawn into these debates in early 2009, after the publication of the two letters from Oaxaca (and subsequent stirrings among geographers) brought the controversy to the attention of the organization's elected Council. Oaxaca loomed large in its annual meeting in Las Vegas, though (as one might expect from the official record of such a meeting) the tone of the discussion is muted in Council Secretary Craig Colten's (2009)[12] official minutes:

> The Council discussed an item recently brought to the AAG regarding ethics and research practice related to the [AGS] Bowman Expeditions [...]. Although this is not an AAG project, the Council decided to examine the AAG's own ethics statement in light of the issues raised [...]. *Baerwald moved that the executive committee appoint a task force, to be approved by council, to examine the AAG Ethics Statement and make recommendations for modifications to the AAG Ethics Statement for consideration at its fall 2009 meeting. The motion was seconded by Agnew, and passed unanimously.* Council recommended that Alexander Murphy, as one of the authors of the current AAG Ethics Statement, serve on the task force. Council members also suggested including members from the Indigenous Geography Specialty Group [*sic*] and the AAG Scientific Freedom and Responsibility Committee. (p. 23, italics in original)

A ten-person committee was soon created, co-chaired by Alexander Murphy and Mike Goodchild, which promptly revised the AAG's "statement on professional ethics." The new text was endorsed by the Council on November 1, 2009 (AAG Council 2009) and posted on the AAG website.[13]

DOI: 10.1057/9781137301758

The revised statement is problematic in numerous respects—McSweeney (2010) provides a visceral critique—but for the purposes of my argument, two sets of criticisms are fundamental. The first concerns the substantive limits of the revised text, which touches on a range of issues—from the need for collegiality and respect for ecosystems, to the scourge of "self-plagiarism"—albeit without substantive discussion of any. In short, the text does not provide meaningful ethical guidelines for examining one's professional practices.[14] We could debate whether it is desirable to define ethical guidelines specific to a given academic discipline (I would argue to the contrary). Nevertheless, we should recognize that the AAG Council's only substantive response to the Oaxaca controversy was to form this committee to revise the ethics statement, and we might expect such an act to have some effect—specifically, to provide a meaningful guide to future action. By my reading the text fails in this respect (the same conclusion is drawn by McSweeney (2010) and Boyce and Cash (forthcoming)). The Council therefore adopted an ineffective and incoherent response to the Oaxaca controversy.

Consider the revisions. The committee merely added some trivial language in new subsections on indigenous peoples (§V.C)[15] and geospatial technologies (§V.D).[16] Read alongside the three prominent elements of the discourse on the controversy, what jumps out is that the third element of the discourse on the Oaxaca controversy—military involvement in and funding of geographical research—*is never mentioned in the revised ethics statement.* Search the ethics statement for the keywords "military" and "Army": there are no hits. Nor does the revised ethics statement indicate *why* it makes no mention of such contentious matters—although they were keenly debated among members of the committee. In a public commentary, committee co-chair Michael Goodchild (2010) explained that the committee had a "lengthy and difficult discussion" about whether to address military funding and collaboration by geographers and concluded that they could not do so. Until the members of the committee choose to explain this decision, we have little basis to analyze the matter. But one thing is clear. Given that this committee was established to examine a controversy that—at least in the texts from Oaxaca and the public letter signed by the 26 geographers—centers on the role of the US Army, it is difficult to see their decision to exclude comment on military involvement except as avoidance or repression.[17]

The second set of criticisms concerns the *process* through which the statement was revised. As its meeting minutes make clear, the AAG

DOI: 10.1057/9781137301758

Council called for revisions in direct response to the Bowman expeditions. We might therefore expect that the committee would discuss what happened in Oaxaca, or to explain why the controversy led the members of the committee to see a need for particularly revisions to our professional code of ethics. They do not. There is no mention of the Bowman expeditions. Indeed there is no explanation or justification for the revisions. They simply happened.

We could go further with this line of critique. As we have seen, the AAG Council appointed a group to revise its professional ethics statement. The group that changed the text was appointed, not elected; its purview was never publicly explained; its work occurred essentially in secret (I know of no minutes of their meetings); and there was neither public discussion nor an explanation of the changes. Whatever we might say about such activities, they bear little relation to the provocation of ethics. The revised statement is like Poe's purloined letter: what explains the silent release of a call for public engagement if not the implicit recognition that the best way to conceal something—the Oaxaca controversy and the AAG's ineffectual response—is to hide it in plain view? To put this question otherwise: what could be a more effective means to dampen the discussion about military involvement in geographical research than to subject the controversy to ethical regulation? Dissent would then be annulled through disciplinary reform.

§3.3.   Many people saw it as fortuitous that, at the time the Oaxaca controversy broke, the Presidency of the AAG was held by Dr. John Agnew, a highly regarded political geographer who has published a body of research on geographical thought, the US military, and hegemony. Agnew led the discussion of the AAG Council that (as noted by Secretary Colten) gave rise to the ad hoc committee and also participated in a number of private discussions and emails about the controversy; some geographers (myself included) speculated that he held strong views about the controversy and would speak out before the end of his presidency. He did so, in a fashion, by devoting two sentences to it in his final "President's Column" editorial in the *AAG Newsletter*:

> [Consider] the controversy that has erupted over research with indigenous communities in Mexico and elsewhere designed to integrate local indigenous land claims with remotely sensed information. Some local groups have objected to the research on the ground that it is "sponsored" by the US

DOI: 10.1057/9781137301758

military, thereby plausibly suggesting that it has anything but the welfare of the local communities at heart. (2009, p. 3)

This passage seems to take us toward a critique of the role of the US military in geographical research, and thus repeat Bryan's argument; but with this opaque remark, the text's discussion of the Bowman expeditions abruptly ends. It simply states that Bowman expeditions (which are not named) were "designed to integrate local indigenous land claims with remotely sensed information" (thus repeating the narrative that the project aimed to help the "local indigenous" people), then notes that this provoked "controversy" since the objections raised by "some local groups" were "plausibl[e]." The tone here (and in Agnew's essay more generally) is one of vague suggestiveness. It seems that the essay's purpose is to downplay the capacity of the AAG (and its president) to speak out on such matters. However limited or imprecise, this was Agnew's position at the end of his presidency in 2009.[18]

In 2010 the ex-president Agnew published a heated response to Joe Bryan (2010), also published in *Political Geography* (on which Agnew happens to serve as an Editorial Board member), entitled "Ethics or militarism? The role of the AAG in what was originally a dispute over informed consent." The gist of Agnew's argument in the essay is that those who have criticized the AAG (during Agnew's leadership of the organization) for failing to act in response to the controversy have unfairly shifted the critique of the Bowman expeditions, first focusing on informed consent and only later raising the military issue. Agnew explains:

> The issue on the table from the start as articulated by all those involved in the controversy was that of informed consent. Never did the involvement of the US Department of Defense in funding the research come up directly at the time in any of the outside complaints, though it was indeed raised, I should add, by some of us on the AAG Council when discussing the entire affair at the spring 2009 Council Meeting. (2010, p. 422)

These statements imply that informed consent could have been cleanly separated from the question of military involvement (a position contradicted by Bryan and Wainwright 2009). Moreover Agnew suggests that *he*, not Bryan, initially raised the matter of military involvement during the 2009 AAG Council meeting ("the involvement of the US Department of Defense [...] was indeed raised, I should add, by some of us on the AAG

DOI: 10.1057/9781137301758

Council").[19] However we may read these propositions, Agnew's text then attacks Bryan:

> By now changing the question under purview, from that of the subject of informed consent [theme one], raised originally and among others by Bryan himself, to that of the militarism implied by any research project with funding from military sources [theme three], and then attempting to indict the AAG as somehow by extension a party to militarism because of its failure to indict, investigate, and presumably potentially ban from membership those originally judged guilty (as yet without any investigation, of course), the very ground on which the original dispute was played out has moved under our feet. This must be a rhetorical device to rope in the entire "geographic establishment" under the charge of militarism. It certainly bears no relationship to what was actually at issue last year. It is a false claim about the role of the AAG in the dispute that I find profoundly disturbing. (2010, p. 422)

This passage is sufficiently hyperbolic to require some patient dissection. Consider the first long sentence. To my knowledge, no one has called for anyone to be indicted[20] or banned from AAG membership. Likewise the parenthetical comment that such punishment should be meted "without any investigation, of course" is remarkable in an essay targeting Bryan, who wrote a letter calling for an "investigation" (Bryan and Wainwright 2009, p. 2). But perhaps the most serious distortion in this first sentence concerns the *temporality* of the controversy. Agnew argues that Bryan (and perhaps others) have "now" shifted the "ground on which the original dispute was played out," implying that what *was* a dispute about consent is *now* a dispute about militarism. Again this implies that the two elements of the discourse can be clearly separated. What if they were united in the two earlier letters from Oaxaca, as they seem to be? But let this pass. Let's suppose that Agnew is right, and there *was* a shift in emphasis in the discourse on the Oaxaca controversy—from [1] consent to [3] military involvement. Well, so what? What is the harm to such a shift in a roiling, open, disciplinary *polemos*? Would such a shift amount to a "false claim about the role of the AAG"? (And is this "false claim" in fact what Agnew "find[s] profoundly disturbing," or is it "the dispute" itself?) The closest that Agnew's text comes to answering these questions is in the second sentence cited above, where it is implied that Bryan (and perhaps others) are manipulating a "rhetorical device" in order to bind ("rope in") people like Agnew ("the entire 'geographic establishment'") with the "charge of militarism." Allow me to restate Agnew's claim, as I

DOI: 10.1057/9781137301758

understand it, more clearly: he is "profoundly disturb[ed]" that Bryan contends that, while serving as president of the AAG, Agnew failed to take a position against militarism in the discipline, whereas, in fact, within the limits of his authority, Agnew did address the ethical dimensions of the Bowman expeditions debate.

Two details should be noted about this intensely argued text, details which suggest that Agnew is projecting his own feelings of regret or disappointment onto Bryan. First, consider the expression "geographic establishment." Although Agnew places these words in single quotation marks (or "scare quotes"), they never appear in Bryan's text. Nor would it be fair to suggest that either of Bryan's texts imply the existence of such a "geographic establishment." Whatever we make of Bryan's arguments, they do not require the existence of such an establishment. Why then does this noun appear in Agnew's text? I suggest that it reflects an anxiety of being seen as a member of such a group (hence the false attribution of this expression via mis-citation is an instance of parapraxis). Second, note that Agnew repeats his claim that the argument about military funding "bears no relationship to what was actually at issue last year," i.e., consent. But Agnew has just written that *he himself made the connection between military funding and problems with obtaining consent at the 2009 AAG Council meeting.* Could it be any clearer that what Agnew finds "profoundly disturbing" is being called out by junior members of the field for failing to speak out when he could have?

My purpose in these remarks on this brief essay is not to analyze the psyche of Professor Agnew but to read these texts and draw out their lessons for our *polemos.* Perhaps the key point is that the AAG president at the time of the Bowman expeditions wanted to speak out against the self-evident militarism of the discipline, yet did not do so. And here we may discern the effects of disciplinary reason. The text reveals certain limiting conditions of possibility that silence geographers, even powerful ones; that enable certain arguments, yet repress others. I think we can accept that Agnew did what might be expected from the AAG president, given the anxious silence among geographers about the third element of the discourse on the Oaxaca controversy. As Jim Blaut emphasized in his (1969) critique of the discipline, the condemnation of imperialism in geography should be directed at no individual, not even the "geographic establishment" (if it indeed exists). The problem concerns our disciplinary formation as a whole.

DOI: 10.1057/9781137301758

§3.4. Viewed retrospectively, the debate in geography surrounding the Oaxaca controversy has been marked by a number of palpable absences. There has been no formal statement by the AAG, nor any workshops or conferences held to sift through these debates. Until now, no substantive books have been published on the Bowman expeditions. The silence has been largely passive but partly active—the latter because critics of the expeditions have been criticized by powerful members of the profession (as for instance in Agnew's critique of Bryan). Such criticism is especially troublesome when directed by relatively powerful actors at untenured faculty and graduate students.

It is difficult not to conclude that the critique delivered by the two letters from Oaxaca has had little effect on Anglo-American geography. Not only has the US Army increased funding to collect geospatial intelligence ("GEOINT" in Army terms) and to practice what it calls "human geography,"[21] but faculty in many geography departments, including my own, are often encouraged to apply for military funding[22] and pressed to hire "security and intelligence geographers." And for its part, the AAG enthusiastically markets the discipline as a good path for those interested in military careers (see figure 1).

To further my investigation of the disciplinary limits of the response to the Bowman expeditions, it will be useful to have a point of reference outside of geography. It turns out that a cognate discipline, anthropology, has also been rocked by similar debates in recent years. To place geography's response to the Oaxaca controversy in perspective, we should contrast it with the responses by anthropologists to the revelation (made at roughly the same time) of the Human Terrain System (or HTS) program, in which the US Army hired anthropologists to help analyze

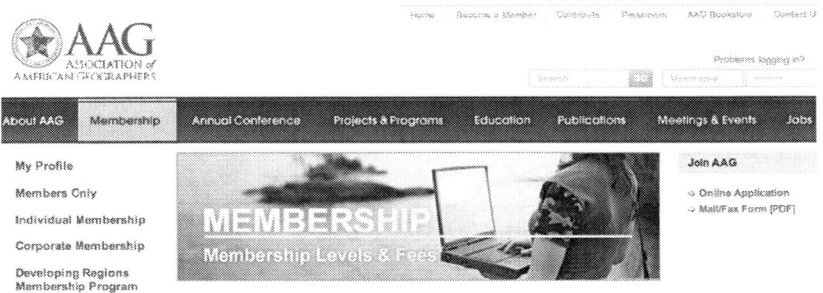

FIGURE 1   *Screenshot from www.aag.org. The figure of a camouflaged militant empiricist geographer, used here to advertise membership in the AAG*

DOI: 10.1057/9781137301758

the "human terrain" in Iraq and Afghanistan (cf. Gusterson 2003; McFate 2005; Kipp et al. 2006; González 2007). The response by anthropologists has been principled, serious, and relatively effective. Anthropology therefore offers a fortuitous and fruitful contrast.

First, compare publications. Academic political disputes play out foremost through texts, and therefore a simple measure of the relative intensity of a debate may be gathered by the volume of substantive intellectual work produced. To my knowledge, as of this writing no human geographers have published a single research article, much less an edited volume or book, examining the Bowman expeditions or the Oaxaca controversy. A number of short statements and essays have been exchanged, but those familiar with academic debates will recognize these as minor works.[23] By contrast, since the mid-2000s anthropologists have published a stack of articles, edited volumes, and monographs in response to the "weaponizing [of] anthropology" (to borrow the title of Price's 2011 book).[24] A thorough review of this literature is beyond the scope of this book, but I would recommend geographers pull down one of the four best in the collection—González (2009); Network of Concerned Anthropologists (2009); Kelly et al. (2010); Price (2011)—that together provide two key contributions to the debates surrounding the Oaxaca controversy. First, they historicize that mysterious subject of the Bowman expeditions' problematic, the "human terrain" (cf. González 2009, pp 25–78; Price 2011, §II). And second, they demonstrate the centrality of Fort Leavenworth, Kansas, in the formation of a post-9/11 academic-military nexus to support a reboot of US Army counterintelligence. Anthropologists have mapped the headwaters of the Oaxaca controversy—its conceptual and institutional sources—across the grasslands of eastern Kansas.[25]

Second, consider institutional responses. The American Anthropological Association (AAA) Executive formed a special committee in 2006 "in response to member criticism of the publication of an ad for CIA employment on the official AAA job site" (AAA 2011), and this committee subsequently advised the AAA Executive on the discipline's engagement with the US intelligence community and national security issues.[26] Partly as a result of this committee's analysis, in 2007 the AAA Executive published an explicit condemnation of the HTS project, arguing that it "raises [...] troubling and urgent ethical issues" for the discipline (AAA 2007). Five such issues are detailed in this text; let us consider two that are pertinent to the Oaxaca controversy. The first problem is that data

DOI: 10.1057/9781137301758

collected by anthropologists could be used in the future by the US military, e.g., for surveillance or targeting:

> [Issue 4] [I]nformation provided by HTS anthropologists could be used to make decisions about identifying and selecting specific populations as targets of US military operations either in the short or long term. Any such use of fieldwork-derived information would violate the stipulations in the AAA Code of Ethics that those studied not be harmed. (§III A, 1)

The use of anthropological data for violent ends could occur without any awareness by those who initially produced the data. Logically, therefore, anthropologists whose data could someday be used by the US military cannot guarantee to their subjects that their research will do no harm. This undermines a key point for the legitimation of anthropological fieldwork.

Their second concern is that the very existence of the HTS program threatens to reshape the identity of the discipline, with potentially violent consequences for anthropologists (even those unaffiliated with the US military):

> [Issue 5] Because HTS identifies anthropology and anthropologists with US military operations, this identification [...] may create serious difficulties for, including grave risks to the personal safety of, many non-HTS anthropologists and the people they study. (Ibid.)

I highlight these two points because they are especially pertinent to the analysis of the Bowman expeditions. Both arguments could be made by geographers—indeed, they were (e.g., Bryan and Wainwright 2009). Yet unlike the Executive Council of the AAA, which promptly addressed these issues through this public condemnation of HTS (among other actions), the AAG Council has to date refused to acknowledge these two fundamental problems.

The AAG-AAA contrast reveals a further gap around the willingness to pose the question of the scholarly legitimacy of research conducted with funding from the US military. Having already published its official criticism of the HTS program shortly after the program became public (2007), the AAA charged its Commission on the Engagement of Anthropology with the US Security and Intelligence Communities (hereafter "the Commission") to analyze the nature and implications of collaboration by anthropologists with the US military. The Commission's report (2009) provides a careful, exacting analysis of the HTS program. It concludes: "where data collection occurs in the context of war, integrated

DOI: 10.1057/9781137301758

into the goals of counterinsurgency, and in a potentially coercive environment [...] *it can no longer be considered a legitimate professional exercise of anthropology*" (AAA 2009, p. 4, my italics). Allow me to repeat this point: the AAA Commission found that wartime data collection with potential military use constitutes unprofessional activity, inconsistent with disciplinary and scholarly norms.

Why, in the face of the militarization of our disciplines, have anthropologists written so much more, and more critically, than have geographers? What led them to act so much more deliberately? I think we can rule out the idea that one discipline was more implicated than the other. The skills involved in mapping "human terrain" are practiced just as much by geographers as by anthropologists. Perhaps it could be claimed that anthropology faced a bigger threat, since the HTS program is larger—both in funding and in personnel—than the Bowman expeditions. Yet the scale of the challenge can provide only part of the answer (and anthropology is a larger discipline, so there is a scale effect). Another answer concerns self-interest. As Gusterson (2009) writes apropos the AAA's actions:

> [T]here was a certain measure of disciplinary self-interest [in the AAA response]: anthropologists can only do research as long as there are people ("informants") willing to talk to them. If they [anthropologists] are suspected of working for the CIA or other organs of the national security state, the ranks of potential interlocutors will thin out, and the most interesting research subjects may be the first to disappear. (p. 49)

This point is important, but it should apply to geographers too, so it cannot explain the divergence.

A stronger answer, I suggest, concerns disciplinary politics. Over the past 30 years, anthropologists have been forced to confront, on one hand, the living legacies of the discipline's imperial roots by postcolonial scholars (see Said 1979; Chatterjee 1986; Spivak 1988; Said 1988; Ismail 2005). Meanwhile, on the other hand, the discipline has been divided by fierce debates over the employment of anthropologists (and the use of their public data) by the US military—particularly in SE Asia during the Vietnam era.[27] The discipline has, in my view, considerable distance to go to address these issues.[28] Nevertheless, the recent debates have undoubtedly politicized the field and led a number of avowedly left anthropologists to organize themselves, which has helped accelerate and intensify the discipline's formal response to the Army's HTS program. On this reading, the AAA's relatively principled and high-profile response is an

DOI: 10.1057/9781137301758

effect of path dependency in disciplinary organizing. Anthropology was long ago sent to war and it is this war, Price wryly notes (2011, p. 11), that "gives anthropology ethics."

Yet was geography not also sent to war? Indeed. A considerable literature documents the long-standing ties between state/military projects and the development of geographical thought.[29] But perhaps the character of the US military's use of geographical thought has changed in ways that we have been slow to detect, leaving us disorganized amidst *polemos*. To make this argument will require us to resituate the entire debate over the Oaxaca controversy on the horizon of the US military's counterinsurgency strategy since September 11, 2001.

Notes

1   Notable exceptions, apart from the previously cited critics of the expeditions, include Gregory (2008), Kearns (2010), and Morin (2011). In fairness to these authors, the "Oaxaca controversy" remains marginal in these texts.
2   Edward Said distinguishes professionalism from expertise, arguing that expertise "has rather little, strictly speaking, to do with knowledge" (1994, p. 79). His example here—fittingly for an essentially anarchist critique of expertise as a form of power—is Noam Chomsky's outstanding work on US wars in South-East Asia (Chomsky is an expert in linguistics, not political science).

    Said's attempt to fend for a committed amateurism has been criticized as romantic, naïve, and—perhaps most damning—inattentive to class and status. These are genuine issues that I cannot address here, though I contend that Said's critique of expertise is worth fending for and his own "amateurism" remains a singular example (Wainwright 2005; on the geographer as anti-expert, see Heyman 2009). I might also mention that I occasionally play with a band called "the Amateurs," which may have influenced these reflections. I acknowledge here the presence and musicking of Bilil Sunni Ali and Oscar Burke.
3   Gramsci's critique of "traditional intellectuals" is one of the most significant statements of its kind in the Marxist tradition. Unfortunately the edition of the prison notebooks used by most Anglophone scholars inexplicably forges one note, "The formation of the intellectuals" (1971, pp. 5–14), by splicing and recombining parts of two separate notes (Q12§1 and Q12§3). Elements of Q12§1 appear separately on pp. 17–33. I thank Marcus Green for sharing his meticulous concordance tables as well as his assistance in sorting through several passages from the *Prison notebooks*.

DOI: 10.1057/9781137301758

4 In my experience most geographers refer to these matters by these two words, "Oaxaca controversy." To work within the prevailing discourse, I will repeat them without scare quotes (after all, there is no true meta-language beyond our *polemos*). Yet we should recognize that the term itself is problematic. *Oaxaca*, the geographical descriptor, distances the matter away from the United States, as if this controversy is remote: something happened somewhere out there, not in Kansas. And *controversy* suggests something distasteful, an ugly spat. Does such language not reflect the polite repression of professional courtesy? After all, in a pair of formal texts the indigenous communities in Oaxaca have raised our eyes to *geopiracy*, of spying by geographers working with the US military: why have geographers in the United States hesitated to repeat this language? Is it not because to do so would be polemical, political, or unprofessional? But "Oaxaca controversy" is a no less political term; it expresses a politics of professional neutrality. The colloquialism forecloses access to the epistemic violence in view from the vantage of the *Rincón*.

5 On the IRB, see chapter 2, note 4.

6 I was invited by the conference organizers to participate in this event but declined. For the conference website, see http://web.ku.edu/~kugso/whgc/.

7 Mutersbaugh, who has worked for more than 20 years in Oaxaca, was one of the first geographers to speak out against the *México Indígena* project. A lengthy email he wrote with a critical discussion of the project (including part of his appeal for a geo-data protocol) was widely circulated on the internet in February 28, 2009. He also criticized Peter Herlihy publicly at the 2010 AAG for failing to return to Oaxaca to discuss the controversy with the communities (Mutersbaugh 2010).

8 The authors define geoslavery as "a practice in which one entity, the master, coercively or surreptitiously monitors and exerts control over the physical location of another" (2003, p. 47).

9 See Bryan and Wainwright (2009) for signatories; count includes the two authors. The Indigenous Peoples Specialty Group also endorsed this letter: "the IPSG membership has endorsed the letter by Joe Bryan and Joel Wainwright, calling for an AAG inquiry into the Bowman Expeditions' México Indígena project, and particularly into the central role of the U.S. Army's Foreign Military Studies Office" (2009a, p. 2).

10 On Radiance Technologies, see www.radiancetech.com.

11 Bryan extends these claims in Wainwright and Bryan (2009) and in Steinberg et al. (2011). I expect some of these arguments will be elaborated in Bryan's forthcoming genealogy of US military involvement in indigenous mapping (coauthored with Denis Wood).

12 AAG Secretary Craig Colten is Carl O. Sauer Professor at the Department of Geography and Anthropology of Louisiana State University and the Editor of *Geographical review*, the in-house journal of the American Geographical Society.

DOI: 10.1057/9781137301758

13   The other members of the ad hoc task force were John Adams, Sarah Elwood, Daniel Griffith, Keith Henderson, Douglas Herman, Sallie Marston, and Olen Matthews.

14   In fairness, the preamble notes that the diverse membership of the AAG faces a "wide variety of ethical considerations" and the purpose of the text is only to generate "discussion and reflection" on some of these. But I doubt it succeeds even in this minimal aim.

15   The new subsection on indigenous peoples comprises only one paragraph. It begins by explaining that research "with indigenous peoples [...] raises special challenges and requires special care." This claim is reasonable—who could deny that colonialism has wrought enormous problems for geographical research among indigenous people?—but the elaboration on these matters is at best vague and at worst simplistic and essentialist. What, e.g., defines the "special care" with which geographers should approach indigenous peoples? Should we be extra polite? What is it that makes indigenous peoples so "special" anyway? Are they more deserving of "mutual benefit" from research than non-indigenous peoples? There are weak and strong answers to such questions; paragraph §V.C offers neither.

16   The core argument of §V.D. is that geospatial technologies "introduce special challenges with respect to potential violations of privacy and confidentiality of individuals and groups. In using these technologies, researchers should make reasonable efforts to protect the health, well-being, and privacy of research subjects."

17   Professor Alec Murphy, co-Chair of this committee, concluded his public remarks at the 2010 AAG by encouraging geographers *not* to emphasize the military when discussing ethical issues. In this regard his public commentary repeated the exclusion of reference to the military during the revision of the ethics statement by the committee he chaired. The December 2006 issue of *Ubique* (the AGS newsletter) notes that Murphy participated in the planning meetings of the Bowman expeditions at Fort Leavenworth (AGS 2006, p. 1). With all due respect to Professor Murphy (whose professional credentials are beyond question), I cannot help but wonder if it was a good idea for the AAG to appoint someone who was involved in the Bowman expeditions—even at "arm's length"—as chair of a committee created in response to controversy caused by those very expeditions.

18   I should note that I exchanged emails with Dr. Agnew and discussed these matters with him in 2009. Our exchange ended after I criticized his hunt for Osama Bin Laden (see Gillespie et al. 2009)—an instance of militant empiricism that Agnew said was intended to be ironic.

19   Similarly, at a point in the text where Agnew directly contrasts his own views with those held by Bryan, Agnew emphasizes that he "probably [has] many more problems with much of what goes for the geographical fieldwork

DOI: 10.1057/9781137301758

tradition than does Joe Bryan" (p. 1). In other words, Agnew *cannot agree with Bryan enough*. Yet he does not specify his problems with the "geographical fieldwork tradition."

20 This may depend on the meaning of "indict," a verb that is curiously doubled in the long sentence under discussion ("attempting to *indict* the AAG [...] because of its failure to *indict*"). The confusion here may stem from the legal *and* extra-legal entanglements of this sign, with attendant etymological complexities (as the *OED* notes, "the history of the Anglo-Norman and Middle English word [indict, *v.*] is not clear," and the links between "the medieval Latin *indictāre*" and contemporary uses have not been resolved). Perhaps the doubling is, to be more precise, solicited by the sign "indict" itself, which conjoins two meanings which are typically suspended in the modern conception of law (if not of justice), viz: "to declare publicly" and "to denounce" or "to charge." In other words, Agnew's text could be read as suggesting that "Bryan has *publicly declared* that the AAG has failed to *denounce* militarism"; yet, equally, Agnew's text could be read as claiming that "Bryan has *denounced* the AAG for its failure to *bring charges* against the Bowman expeditions." This does not exhaust the possibilities, of course, and the resonances of the potential permutations far exceed Agnew's intentions.

And what of the texts from Oaxaca? To return to the *OED*: "The sense of *endite, indict*, may have arisen from Latin *indīcĕre* "to declare publicly," taken as in Italian *indicere* "to denounce" (Florio); but it comes near to a sense of Latin *indicāre* to indicate *adj.*, "to give evidence against"; and it appears as if there had been, in late Latin or Romanic, some confusion of the Latin verbs *indicāre, indīcĕre, indictāre*." Perhaps we could say that *the texts from Oaxaca indict* (publicly declare, give evidence of) *geopiracy*. Insofar as that is the argument at the heart of this book, let me add the clarification that they do not make this indictment within or before the law. Our reading of them might aim to bring about another calculation of justice (Derrida 1990), an *other indictment*.

21 To my knowledge, human geographers lack a publicly accessible text that reviews all of the US military's work in and on human geography. However there is abundant evidence that it is massive. Today geographical research (as commonly understood) for the US military is coordinated by the National Geospatial Intelligence Agency (NGA), which is

the nation's primary source of geospatial intelligence [...] for the Department of Defense and the U.S. Intelligence Community. As a DOD combat support agency [...] NGA provides GEOINT, in support of U.S. national security and defense, as well as disaster relief. GEOINT is the exploitation and analysis of imagery and geospatial information that describes, assesses and visually depicts physical features and geographically referenced activities on

DOI: 10.1057/9781137301758

the Earth. [...] NGA employs approximately 16,000 government civilians, military members and contractors[.] [...] NGA's budget is classified. (https://www1.nga.mil/ABOUT/FAQS/Pages/default.aspx, accessed March 16, 2012)

With "around 16,000" employees, the NGA's employees outnumber the human geographers presently employed in US universities. While its budget is secret, it is fair to surmise that the NGA is very well resourced. The NGA manages its own GEOINT school, the National Geospatial-Intelligence College (NGIC), which is likely the largest center for geographical education in the United States. According to the NGA's website:

Each year, NGC trains more than 15,000 students—including government civilians, military members and contractors—from across NGA, the Department of Defense (DOD), U.S. Intelligence Community (IC), as well as federal, state and local governments and foreign mission partners.

An unclassified presentation by Steven Grant (2011) provides a glimpse of recent US Army geography programs. Grant notes: "The US Army has developed a coherent research program to provide the required tools to acquire, interpret, and communicate [social, geographical, and cultural] information [...] which is almost always communicated geospatially." He adds that the US Army's "Human Geography Working Group, led by NGA, should allow socio-cultural information to be shared within a coalition environment."

The theme of the September/October 2012 issue of the National Geospatial-Intelligence Agency's magazine is "human geography." There we read that the NGA has collaborated in creating a "H[uman] G[eography] Senior Steering Group [...] composed of 22 Department of Defense and intelligence community agencies"; also that the NGA, in collaboration with the US State Department, "has established a Worldwide H[uman] G[eography] Data Working Group to build voluntary partnerships around human geography data and mapping" (2012, p. 14).

22   For instance, we are encouraged to apply for the MINERVA grants—funded by the Department of Defense and processed by the NSF—which support the analysis of "the social, cultural, and political dynamics that shape regions of strategic interest around the world" (http://minerva.dtic.mil/index.html, accessed April 12, 2012). The SSRC (2008) has posted a collection of essays on the Minerva controversy at http://www.ssrc.org/programs/minerva-controversy/.

23   I include my own letters (Bryan and Wainwright 2009; Wainwright and Bryan 2009b) and lectures in this critique.

24   In addition, anthropologists have organized numerous workshops and conference sessions, built detailed websites, written scores of journalistic reports and essays, and more. See for instance John Stanton's writings at

DOI: 10.1057/9781137301758

http://zeroanthropology.net and the annotated bibliography on HTS and Minerva at the website Culture Matters at http://culturematters.wordpress.com/2008/08/21/annotated-bibliography-on-hts-minerva-and-prisp/.

25 The FMSO is based in Fort Leavenworth. On the Leavenworth nexus, see González (2009, pp. 35, 44–49), NCA (2009, pp. 88, 101, 107), and Price (2011, pp. 35–37, 98–103; 133–135; 155–172). Much as I appreciate this literature, it often seems to suggest that the militarization of the social sciences is principally a problem for anthropology. It is not. The problems we face exceed our disciplinary boundaries and we cannot defeat militarism by defending our disciplines against the Army. Questioning disciplinary formations is essential to our struggle against the weaponization of social analysis.

26 See Marcus (2009) on the creation and politics of this body. Reading Marcus, it seems that this committee was created (appointed) by the AAA's executive leadership.

27 See NCA (2009, chapter 1), González (2009, chapter 1), and Gusterson (2003).

28 Whether the recent period is exceptional is an open question. A strong argument can be made that the discipline of anthropology has always been shaped by empire (see Ismail 2005).

29 The literature by geographers on empire, geographical knowledge, military science, and cartography is considerable. See especially Blaut (1969), Smith (1994), Driver (2001), Cloud (2002), Smith (2003), Sundberg (2003), Woodward (2005), Barnes (2006 and 2008), Gregory (2008), Wainwright (2008), and Kearns (2010).

DOI: 10.1057/9781137301758

4
Geography Counterinsurgent

Abstract: *Conrad on geography militant ~ Barnes on geography and the mangle of war ~ The birth of geography counterinsurgent ~ US failure to win hegemony in Iraq ~ A shift in US counterinsurgency strategy ~ Petraeus's influence in the Bowman expeditions ~ "Human terrain" defined ~ The need to re-tool old-fashioned geography with GIS ~ Geoproperty and geosecurity ~ An aside on the scholarly merits of the Bowman expeditions ~ Relations between US militarism and Anglo-American geographical thought summarized.*

Keywords: Fort Leavenworth; geography counterinsurgent; geoproperty; geosecurity; GIS; hegemony; mapping human terrain; macroscope; David Petraeus; US Army *Counterinsurgency Field Manual*

Wainwright, Joel. *Geopiracy: Oaxaca, Militant Empiricism, and Geographical Thought*. New York: Palgrave Macmillan, 2013. DOI: 10.1057/9781137301758.

DOI: 10.1057/9781137301758

[I]t would be fatuous to ignore the effect of money and worldly ambition on scholarly writing and research. Many professors are subject to these pressures, and it is quite possible that the resulting dangers pose a greater risk to scholarship than any threats arising from conventional attacks on academic freedom.

Bok (1982, p. 26)

Jerry Dobson conceived the [Bowman expeditions] because he [...] was troubled over US intelligence failures [...]. "What's missing[, Dobson argued,] is open source[1] geography[...] based on the type of fieldwork and data analyses that geographers do routinely in every region on earth. I firmly believe the only remedy is to bring geography back to its rightful place in higher education, science policy, and public policy circles." [...] [G]ood old fashioned regional geography can be re-tooled with digital technologies and humanistic methodologies. Dobson's notion was embraced and supported by the Foreign Military Studies Office (FMSO) in nearby Fort Leavenworth.

Herlihy et al. (2006, p. 1)

§4.1   In one of his final essays, Joseph Conrad divides the history of geography into several phases. After a long period of "fabulous geography" in which "circumstantially extravagant speculation" ran wild and maps were filled with imaginary spaces, a new sort of science emerged in which men (Conrad names Columbus, Tasman, van Diemen, Cook, and Livingston) remade the world through "adventurous action" and courageous expeditions—the sort that makes boots muddy and generates field data. Such were the days of "geography militant." I write "were" because Conrad's discussion of this era is nostalgic, backward-looking, reflecting upon a passing era. By the early twentieth century (Conrad wrote this essay in 1924), the unmapped world was disappearing, the blank spaces on the mapped filled in.[2] You can sense this tinge of nostalgia in Herlihy et al.'s call to bring back "good old fashioned regional geography"—see epigram—while noting its need to be "re-tooled."

Conrad's remarks on the twilight of geography militant bring to mind the most important US geographer of that era: Isaiah Bowman. Bowman's often contradictory contributions to geographic thought and practice are carefully analyzed in Neil Smith's critical biography (2003),

DOI: 10.1057/9781137301758

which scrutinizes the remaking of the geographies of American empire through the lens of Bowman's career. To summarize briefly, Bowman put geography at the service of US empire, which he saw would require building US-led *global* institutions that could regulate a liberal political-economic order conducive to US interests.[3] Bowman thus labored to produce "a modern, liberal, American geography of the world," one where the US rule was simply "beyond politics" and rooted in the organization of space (2003, p. 187). Among other things, this work included Bowman's attempt to resolve the "vital ethnographic problems of Europe"[4] by creating a cartographic database with which to redraw Europe's boundaries at the 1919 Paris Peace Conference. In this and other respects, the Bowman expeditions are aptly named (if atavistically so), since like Bowman these expeditions emerge in an effort to put geography—exploration, measurement, mapping—to work to meet the needs of the US empire.[5]

By Conrad's estimation the era of "geography militant" would be replaced by an effete era of "geography triumphant," where the expansion of mass travel offered us only the lazy comfort of knowing the already-known world. Things did not turn out quite so. By the mid-twentieth century, Anglo-American scholarly geography had sunk into "geography's underworld," to use Trevor Barnes' felicitous expression (2008). In a quartet of carefully argued papers Barnes (2006, 2008; Barnes and Farish 2006; Crampton and Barnes 2011) elaborates upon the ways that the discipline was rended by the "mangle" of World War II:

> During the earlier period of Empire [roughly, the eighteenth and nineteenth centuries], the military were interested in geography because of the traditional skills that the discipline brought—the drawing and interpreting of maps, the understanding of topographic features, the knowledge of regions and their boundaries. Geographers did what they always did, and while they might have come out of imperial war as changed individuals, the discipline they practised did not significantly alter. But this was not true in the later period [i.e., the 1950s and 1960s]. The discipline's very centre was displaced, shifted and realigned, mangled, in part because of its connections to the military, its connections to the Cold War. This fact has not been usually recognised. It is one of the discipline's secrets. [...] [G]eography does have its secrets, does have its underworld. (2008, p. 15)[6]

Analyses of the role of the state/military in shaping Anglo-American geographical thought tends, unfortunately, to be overly historicist and empiricist, organized around specific chapters in the history of the discipline

DOI: 10.1057/9781137301758

(I cite Barnes's work in part because it runs against this grain). We cannot presume continuity between the "geography" of, say, the 1920s, the 1960s, and today: the discipline itself must be kept in question. With this in mind we should ask a pertinent question: If Conrad and Bowman were writing at the dusk of "geography militant," and the Cold War era was one of "geography's underworld," what can we make of our own time? What is geography's present? The Oaxaca controversy demands that we ask these questions anew, re-evaluate the role of the state/military in geography, and try to discern whether there is anything new coming out of the mangle. Perhaps geography's "underworld" has come out from below? Perhaps the expeditions simply repeat the ideology of Bowman, as their name suggests? Perhaps they harken a return to geography militant?

These questions are taken up below but, for the sake of clarity, let me offer these brief answers. I do not think geography today reflects Bowman *redux*, though certain parallels are noteworthy. Nor are we seeing the rebirth of "the underworld." A stronger case can be made that we are witnessing a return, in new guise, to late nineteenth-century geography militant. Yet as always, historical repetition implies difference.[7] Where the Bowman expeditions revive the spirit of Bowman, today the militarization of empirically-driven geographical fieldwork has returned in a more "open" and less "elite" form, where team-based data collection and GIS-coordinated analysis serve to calibrate counterinsurgency. Both the repetition and difference have, therefore, technosocial aspects. To understand them we must briefly consider the state of US hegemony today—fraying amidst global crisis, yet confidently jittery at the dawn of an era of drone warfare.[8] My thesis is simple. We are entering an era of *geography counterinsurgent.*

§4.2 Let us begin again, not in Oaxaca but in Iraq. To briefly recapitulate the "known knowns": Iraq never had any Weapons of Mass Destruction (WMDs), hence the war's justification was bogus; the US invasion occurred without the endorsement of the UN and in contradiction of international law; hundreds of thousands of Iraqi civilians were killed; the resulting nation-state is at best ungovernable and may well collapse into bloody civil war; and the financial costs of the war have escalated into the trillions of dollars. We could go on but a full accounting is ultimately impossible—and in any event it is too soon to tally the toll since the war on Iraq has not yet come to a discernible end.

For that matter, we could debate exactly when the war began. Let us recall that the United States was no neutral observer of the Iran-Iraq war

DOI: 10.1057/9781137301758

(1980–1988), invaded Iraq in 1991, bombed the country and crushed its economy with sanctions throughout the Clinton years—and then the Bush White House began planning its invasion of Iraq immediately after September 11, 2001.[9] Successive waves of military-politico-economic assault have thus evolved over a quarter century. Under the aegis of the Global War on Terror, the US state framed Iraq and Afghanistan within a specific geographical imagination, one where the instabilities of a dangerous and anarchic world could be bisected—with us or against us, good versus evil—upon an axis spinning around Baghdad (Gregory 2004). The invasion of Iraq was certified by the need to destroy WMDs; this mission having ended for their awkward nonexistence, the US military and political strategy in Iraq effectively became *counterinsurgency*, much as it did in Afghanistan after US forces swept across the Pakistan's northern highlands. The renewal of interest in counterinsurgency strategy among US military strategists results, therefore, from the errant charges of a declining imperial power.[10]

The key text with which to appreciate both the continuity and novelty of the US military's theoretical take on counterinsurgency is the US Army's *Counterinsurgency Field Manual* (2006; hereafter the *Manual*). Written in haste at Fort Leavenworth by a team of military scholars in 2005–2006—"to fill a doctrinal gap" (Petraeus and Amos 2006, p. xlv)—mainstream commentators celebrate the *Manual* as a "radical" foundation for a novel counterinsurgency (or COIN) strategy.[11] In fact these principles of COIN strategy described in the *Manual* are generic to imperial struggles for hegemony; as Sewall notes in her introduction to the *Manual*, British counterinsurgency campaigns in colonial Malaya and Kenya were based upon the same considerations (2006, p. xxiv)— three in particular. The first is that the purpose of counterinsurgency is to win *consent* (see Nagl et al. 2006, §§1-3, I-4, I-14, I-40, I-79, I-153, and *passim*). Hence, victory in counterinsurgency "is achieved when the populace consents to the government's legitimacy" (2006, p. 6). Second, understanding the enemy and winning consent both require a keen awareness of *geography and culture* (2006, §§1-80, I-84, I-125, I-126, B5-9, and *passim*). This is because "the interconnected, politico-military nature of insurgency and COIN requires immersion in the people and their lives to achieve victory" (§1-125, p. 40). And third, successful COIN requires an absolute unity of effort across military and nonmilitary, state and civilian lines (§§1-40, I-136, chapter 2 *passim*). The role of the civilian is to support the military. Sewall is probably right to suggest that the 2006

DOI: 10.1057/9781137301758

counterinsurgency doctrine's "most important insight" is that, in fighting its wars today, the United States "must align its ethical principles with the nation's strategic requirements" (p. xxii). Yet Sewall does not specify how we are to discern these "strategic requirements," and seems to presume that US citizens will consent to aligning our ethics to meet them.

The question of the realization of these concepts brings us to David Petraeus, who supervised the completion of the *Manual* in Fort Leavenworth after his two "tours" in Iraq.[12] Patrick Cockburn summarizes the importance of this particular organic intellectual:

> The great success of General David Petraeus as US commander in Iraq was to persuade many Americans that they had won [in Iraq] when they had not. He also convinced them that the war had ended, when it had not [...]. In practice, the verdict of Iraq is likely to hang over US foreign policy for a long time to come. The war may not have had a clear winner, but it showed that *superior military force no longer easily translates into political victory.* (Cockburn 2011, no page, my italics; contrast Petraeus 2007)[13]

Cockburn reminds us of a conclusion reached long ago by Gramsci, namely that hegemony cannot result from mere coercion. This is a point that Petraeus probably appreciates as well as anyone. It was Petraeus who was asked to lead the United States to victory in Iraq after the initial bombardment and conquering of Iraq's territory, once the US military had demonstrated its incapability of overcoming the waves of resistance to US occupation (particularly within the Sunni triangle). Petraeus— General of the surge, intellectual author of the "new" US counterinsurgency strategy, presently director of the CIA—matters enormously here because he also inaugurates the Bowman expeditions.

Let's consider an insightful text that illuminates the connections between geopiracy and Petraeus' historical singularity, a 2008 report by Peter Herlihy and the Bowman expeditions team written to the FMSO (Herlihy, Dobson, Aguilar Robledo, Smith, Kelly, Ramos Viera, and Hilburn, 2008a; hereafter, Herlihy et al. 2008a). The report credits the expeditions to Petraeus' intellectual leadership, who is described with ostentatious fealty ("General David H. Petraeus [...], erudite 21st century soldier-scholar, with an M.P.A. and Ph.D. degrees from Princeton University"). Petraeus is cited by Herlihy et al. as the source of the connection between geography and mapping human terrain:

> [K]nowledge of the cultural "terrain" can be [...] even more important than, knowledge of the geographic terrain. This observation acknowledges

DOI: 10.1057/9781137301758

that the *people are* [...] *the decisive terrain*, and that we must study that terrain in the same way that we have always studied the geographic terrain.
(Petraeus 2006, p. 8, cited in Herlihy et al. 2008a, p. 22, my italics)

What does Petraeus mean, "people are the decisive terrain"? This seems
an empty banality like Woody Hayes' slogan, "you win with people."[14]
Of course people are decisive, you say; how does anything ever change
except by changing people? But Petraeus' comment has a specific political
meaning and it is no secret. Stated in the wake of the disastrous failures
to win hegemony in Afghanistan and Iraq, Petraeus' point is that the
decisive terrain of struggle is not the sum of airspace plus terrain—the
geographic terrain—that is often easily won by the US military. Rather,
the "decisive terrain" upon which the US military wins and loses wars
is found in *the minds of the enemies we must defeat and the consciousness
of those we must conquer.* This is a restatement of Gramsci's insight that
hegemony cannot be created merely through coercion, but requires the
production of "active consent" (Q11§51 1971, p. 370).[15]

Let us return to Herlihy et al. (2008a) resuming where the text comments upon Petraeus' insights on the centrality of mapping human terrain and the origins of *México Indígena*:

General Petraeus met with members of the AGS Council and the *México
Indígena* research team in October 2006, before returning to command U.S.
troops in Iraq.[16] The general detailed his thoughts on military and humani-
tarian work among foreign lands and peoples. AGS President Dobson
related how geography, as the science of people and place, combines the
"cultural" and "geographic" terrains into one "cultural landscape" [...]. [In
response, Petraeus, t]he seasoned pilot and field commander[17] welcomed
the AGS Bowman Expedition concept and he recognized the importance of
doing participatory research in the Mexico project. He spoke frankly telling
the group how U.S. troops were unprepared for these "cultural terrains" and
of the need "to get the troops smarter quicker." There was consensus about
the need for cultural terrain studies and that *geographers were best equipped
to produce this cultural terrain, or what is now being called the "human terrain."*
(Ibid., p. 22, my italics)

"There was consensus..."; "what is now being called the 'human terrain'":
why is this final sentence written in passive voice? What would make
such a muscular, masculine text so flabby and passive? Superficially,
this deadening prose suggests that its authors were imbibing too much
military-base chatter. More substantively, I fear that the passivity of tone

DOI: 10.1057/9781137301758

in this and other key sentences—sites of mutual adequation between geographical empiricism and practical militarism—lack visible subjects and active verbs because the authors of these texts did not wish to leave fingerprints, i.e., did not wish to specify that they were on hand at the assignment of geography's new mission, viz: "to produce this cultural terrain, [. . .] the 'human terrain,'" for the US military. For indeed, who or what could authorize this group of men, gathered at Fort Leavenworth and wearing the new uniforms of geography counterinsurgent, to decide that geographers should map the world qua "human terrain" for the US military? (To decide this by "consensus" no less—apparently there were no conscientious objectors present.)

Perhaps this seems too strong, since, as I noted earlier, most of the texts in the literature produced around the Bowman expeditions refer to "human terrain" in extremely vague terms and, when pressed on this issue, Professors Herlihy and Dobson have denied that there is any connection between the HTS program and the Bowman expeditions. We have already seen that the founding texts of the Bowman expeditions plainly organize their mission around mapping human terrain. But what exactly is to be mapped? What constitutes human terrain? The term should seem a bit strange since we are not accustomed to thinking of human beings as mere "terrain". "Terrain" is a synonym for the Earth's physical landscape, usually conceived as a stratum upon which humanity acts. To equate human beings with this material surface implies flattening, simplifying, and dehumanizing. How is this "human terrain" to be produced?

It is to the credit of Herlihy et al. (2008a) that this text offers, after the just-cited passive sentence, a definition of "human terrain". Notably, the definition does not come from a scholarly text, but from the US Army:

> The "human terrain" is defined by military scholars: "as the human population and society in the operational environment (area of operations) as defined and characterized by sociocultural, anthropologic, and ethnographic data and other non-geophysical information about that human population and society." (Herlihy et al. 2008a, pp. 22–23)[18]

Geographers are henceforth to produce maps of human terrain, which is the sum of "sociocultural, anthropologic, and ethnographic data" about the "human population" in the military's "operational environment." Mapping human terrain means producing a complete and encyclopedic data set for the entire globe, covering every potentially targeted

DOI: 10.1057/9781137301758

population, every would-be enemy. "Human terrain" signifies *everything* concerning humanity: the totality of the world, minus physical geography. Such a concept would be useless, analytically, which may explain its absence in the human geography scholarly literature.[19] Nevertheless, by the logic of the Bowman expeditions, today the purpose of geography is to map human terrain; this task represents culmination of our disciplinary heritage:

> No discipline or profession naturally integrates these data as effectively as geography and the geographer. Indeed, *the human terrain is at the very core of geographic scholarship, being all but synonymous with cultural landscape.* (Ibid., my italics)

Fortuitously, geography is organized around an object, "cultural landscape"—at least in the Sauerian tradition celebrated by the Bowman expeditions (e.g., Herlihy et al. 2006, p. 6)—which is "almost synonymous" with the object of military concern, "human terrain." We need not change objects. We need only realize the purpose of our mission. The US military can help us here, because although it knew it needed help mapping human terrain, it was only when Petraeus sat down with Dobson and company at Fort Leavenworth that the Army realized that what it had needed in Iraq and Afghanistan all along was—human geography:

> [T]he military is clearly interested in geography, but isn't familiar enough with the discipline to recognize it *by that name*. One notable exception, the U.S. Foreign Military Studies Office (FMSO), funded by the US Department of Defense, has been the primary source of financial support for the first AGS Bowman Expedition, but *our purpose is academic not military* [...]. (Herlihy et al. 2008a, pp. 22–23, my italics)

This final claim, that their "purpose is academic not military," would be more reassuring were it not contradicted through the entire text. And if the same authors had not stated that they were "constructing a very broad national-level GIS that the FMSO would find useful in many different types of analyzes [sic]" in *México Indígena* Project Status Report Two (July 2005). And if the same authors had not been "hosted" by military leaders at the FMSO headquarters in Fort Leavenworth (Dobson 2006b) and later reassured their military funders that they were maintaining regular contact with the "FMSO, NGA [National Geospatial-Intelligence Agency], and Marine Core [sic] Intelligence Agency (MCIA) representatives."[20] And if the problematic for this research had not originated because " 'human terrains' demand accurate,

DOI: 10.1057/9781137301758

Ꮜ Ᏼ Ꮖ Ꭴ Ꮜ Ꭼ
N O T E S *from* T H E A M E R I C A N G E O G R A P H I C A L S O C I E T Y

Fort Leavenworth Hosts AGS Council
Jerome E. Dobson, President, AGS, Professor of Geography, University of Kansas

Each year the AGS Council visits an institution where geography is practiced as an integral part of one or more major missions directly affecting science and society.

Always, our purpose is to learn from people engaged in vital work, but also to tell them about AGS and promote geography. This year we chose Fort Leavenworth, Kansas, home of the Command and General Staff College, Foreign Military Studies Office, and other important institutions. Our visit took place on October 20, 2006.

Fort Leavenworth

Fort Leavenworth is the oldest continuously operating

military installation west of the Mississippi River. Indeed, with the impending deactivation of Fort Monroe in Virginia, it will become the second oldest active military post in the United States, surpassed only by the U. S. Military Academy at West Point.

This historic post, noted for its campus setting, open green spaces and hometown character, is the home of the US Army's Combined Arms Center (CAC), a major subordinate headquarters of the US Army Training and Doctrine Command. It has often been called the "Intellectual Center of the Army" and is, in many regards, "home base" for the majority of field grade officers throughout the Army.

(continued on page 2)

Photo by James W. Thomas

"AGS Councilors and friends gathered in front of the statue of the Buffalo Soldier at Fort Leavenworth. From left to right, AGS Councilors Rickie Sanders, Clifton Pannell, John Frazier, and Jerry Dobson; FMSO Director Karl Prinslow; U. of Kansas professor Peter Herlihy (PI for AGS Indigena research project in Mexico); AGS Councilors Frederick Nelson and John E. Gould; and FMSO Senior Research Director Geoffrey Demarest. In front, AGS Executive Director Mary Lynne Bird. Note: AGS Councilors David Keeling and Alexander Murphy joined the group later in the visit."

FIGURE 2   *Cover of* Ubique: *"Ft. Leavenworth Hosts AGS Council"*

on-the-ground field knowledge" (*México Indígena* ca. 2005). And if their conception of "human geography" were not consistently equated with the mapping and analysis of "human terrain" as defined by military scribes.

§4.3   How exactly does one map and analyze human terrain? With which "tool" does old-fashioned regional geography "re-tool" for this end? In a word, the answer is GIS; but the matter is a bit more complex. As "GIScientists" are acutely aware, a Geographical Information System may be the organizing nexus within which the human terrain becomes mapped and analyzed (if not, indeed, produced); but GIS analysis proper is only ever possible because, before the spatial data is analyzed, there are people in the field collecting empirical data, taking notes, sketching maps, annotating metadata, and so on; likewise before the analysis proper commences, there are specific problems and questions in mind that will guide the operation of the GIS. In sum, the way one maps and analyzes human terrain is through planning, conducting fieldwork, collecting data, and organizing the material thematically and analytically, building the dataset within the GIS, conducting the analysis, and so on.

DOI: 10.1057/9781137301758

There is often a tremendous amount of labor involved, all of it social. Governing these social relations are implicit norms, including especially the norm of empiricism that invisibly guides the various actors through the processes of translating the world into the GIS across various technosocial boundaries (linguistic, territorial, conceptual, and so on).

Yet again: why GIS? What is it about GIS that makes human terrain apparent in a way that might help "get [US troops] smarter quicker," to repeat Petraeus' commanding words? The literature on the Bowman expeditions provide two distinct answers to this question. To put the two answers in terms of a well-known dyad of political geography, one answer centers on the spatiality of state/territorial power, the other on capitalist space economy. I begin with the latter:

1. *Geoproperty*: Geoff Demarest, formerly a researcher for the US Army's FMSO and a graduate of the US Army School of the Americas, asks and answers the pertinent question:

> Why the emphasis on geographers and GIS? Because [...] property lies between law and land, land and economics and between military strategy and human rights. Property is the agreement about agreements, the basic conflict resolution mechanism, the social contract, the most successful of civilizing institutions, and the indispensable secret that permits liberty and [...] peace to flow as one. (2009, p. 315)

This is the central thesis of Demarest's book *Geoproperty* (1998) and a 2009 study for the Defense Intelligence Agency, a pair of manuscripts devoted to the proposition, or "the indispensable secret," that peace flows from property rights.[21] In Latin America—a region Demarest has studied for the US state/military in various roles for more than two decades[22]—the principal barriers to the clarification and consolidation of property rights, and hence *pax Americana*, is the continued existence of the commons, i.e., spaces and places where poor people, often indigenous people, manage their world together and without the purported security of private property rights to land. This is what sets Demarest against indigenous rights to community lands, a point that was not lost in Oaxaca, one of the largest regions of commonly managed indigenous lands in the hemisphere. Demarest's concerns with geoproperty are reiterated throughout Batson's (2008) celebration of the value of geographical research for the US military, a study that features the Bowman expeditions as a model case, intended for replication (pp. 35–52).

DOI: 10.1057/9781137301758

I should note that both Batson and Demarest—who were, I believe, employed by the US military at the time—traveled to Oaxaca alongside the Bowman expeditions. According to *México Indígena* Project Status Report Two, Dobson brought Demarest to Mexico in July 2005. Demarest's presence in Oaxaca raises two noteworthy political issues. The first is national: Articles 9, 10, and 32 of the *Constitución Política de los Estados Unidos Mexicanos* prevent the involvement of foreign militaries in Mexico. Mexican activists have noted, therefore, that the Government of Mexico, or at least its military, must have given tacit support to the Bowman expeditions— a charge denied by the Mexican Congress.[23] The second issue concerns Oaxaca. In 2006, while these geographers conducted their expedition to the *Rincón*, Oaxaca was in the midst of a great civil conflict, the "APPO uprising," during which a coalition of radical and progressive social groups drove the reviled, corrupt Governor of Oaxaca from the capital and successfully "occupied" the City of Oaxaca—despite consistent and often deadly attacks by armed police and soldiers—for more than a month.[24] It is no exaggeration to say, therefore, that the Bowman geographers conducted their expedition alongside US military agents, at least for a time, in a war zone.

2. *Geosecurity*: Demarest also elegantly summarizes the second answer to the question, why the emphasis on geographers and GIS for the analysis and mapping of human terrain? What is it that makes GIS "the certain future of intelligence" (Demarest 2009, p. 291)? The answer derives from the analytical capacities that come via GIS integration of disparate types of spatial data. Military strategists highlight the ability of GIS to combine large (e.g., global and national) data sets and rapidly "downscale" spatial analyses to support small groups of soldiers—even in remote or rural areas. The value of such battlefield logistics, at least to the US military, has increased since September 11, 2011. Demarest (2009) explains:

> As a result of the Global War on Terror, the main organization unit [of the US military] seems now to be a special forces team, comprised perhaps of a dozen people. [...] The basic elements of geography that guide our operational intelligence view of the world needs to be changed [...]. Instead of the country, we need to focus on the county and its international cousins. Almost everywhere in the world the best property intelligence [remember that property intelligence is an essential variable for Demarest] is at the level of territoriality and below. (2009, p. 318)[25]

Dobson provides a scintillating view into the mind of a liberal enthusiast for the US state's ownership of a panopticon-like GIS: "It's one

DOI: 10.1057/9781137301758

thing to know where each bomb will fall, and GPS can tell you that. It's quite another to know where the people are, and that requires a GIS" (Dobson 2005b, no page). Very well. Yet we should ask, why must the US government know the precise location of people in, say, the *Rincón de Ixtlán*? One line of answers to this question focuses on the qualities of this particular place: its commonly held lands, potential resources, history of civil resistance, and so on. But the second and more significant answer concerns not Oaxaca but the view of the world from Fort Leavenworth. The Bowman research in Oaxaca, we are often reminded by the project's source texts, was a "*prototype* expedition" (Herlihy et al. 2008; my italics). The Army's aim is not to map the *Rincón*, per se, but the entire world. Oaxaca was only a testing ground.

§4.4   We come now to a matter that should be central to any scholarly analysis, but when raised here may seem slightly awkward: the question of the merits of the results of the Bowman expeditions as research. This will require shifting levels in my critique, from the immanent to the merely disciplinary. This subsection conflicts with the ethos of my presuppositions. Then why take up the question of the intellectual contributions of the expeditions? The answer is simply because any controversial academic research must be examined, in part, for its potential intellectual contributions. The problem is that most such evaluations remain narrowly limited to prevailing norms and conforms to disciplinary reason. My reading here will, of necessity, bend to these norms; then I will re-open the question of the limits of such a critique.

* * *

As we have seen, the participants in the Bowman expeditions claim to conduct serious research. This position might be convincing if they were publishing original scholarly works in good academic journals. Yet, by and large, they are not. To my knowledge, only three articles have resulted from this body of work (Smith et al. 2009; Herlihy et al. 2008; Kelly et al. 2010). One of these (Herlihy et al. 2008) describes the *México Indígena* project as the "prototype" of the Bowman expeditions. This piece, published in the in-house journal of the AGS (sponsor of the Bowman expeditions), simply reprints many of the flowery statements about the project written by the same authors in their reports to the US Army-FMSO. By my reading, this text is essentially a compilation of

DOI: 10.1057/9781137301758

descriptive project documents; it is not so much a research article as an advertisement for research funding.

By contrast the other two papers (Smith et al. 2009 and Kelly et al. 2010) constitute research papers—or at least one research paper that has been parsed into two. Both articles were published in the *Journal of Latin American Geography* (aka *JLAG*),[26] by the same four authors (with an additional two added to the 2010 text) on the same topic (the effects of PROCEDE on indigenous title), using the same data set (the *México Indígena* fieldwork in the Huasteca Potosina and the Sierra Juárez), to make the same argument. Their argument, in short, is that Mexico's neo-liberal land certification program, called PROCEDE, has had differential outcomes and some negative consequences for indigenous communities. Yet this critique is made in a rather schizophrenic fashion, since these texts present PROCEDE as progressive, inevitable, "laudable" (2009, p. 176), even "revolutionary" (p. 198). Reading these papers I get the impression that at least some of the authors are liberal critics of neo-liberal policies who—quite rightly in my view—came to be persuaded that PROCEDE should never have been created. But if they came to this intuition, they stop short of saying so. Rather, we read:

> While the ostensible goals of the program [PROCEDE] are laudable— namely, providing greater land security and stimulating investment and enhancing productivity in the rural sector—our results raise serious concerns, including the generation of new boundary disputes, the breakdown of community institutions, increased socioeconomic differentiation, loss of forest cover, and threats to cultural survival of vulnerable indigenous populations. (Smith et al. 2009, p. 176)[27]

Clearly these critical remarks on PROCEDE collide with the argument (made most vociferously by Demarest, but implied by other texts in the Bowman expeditions: e.g., Herlihy et al. 2006) that the certification and formalization of property rights is necessary for the development ("enhancing productivity"), opening ("stimulating investment"), and pacification of indigenous Mexico.[28] The more pertinent issue here concerns the relationship between these insights on PROCEDE's effects—which mark, incidentally, the critical high-water mark of the entire Bowman *oeuvre*—and the US military's role in funding and shaping these works. Remarkably, neither of the *JLAG* papers includes any self-reflective remarks on their collaboration with the US military or its influence on their research. No doubt this is because the authors have

DOI: 10.1057/9781137301758

worked to convinced themselves that the veracity and value of their research strictly results from their fieldwork, data, and GIS analysis. Such a view betrays the mark of empiricism, of course. It also helps us to appreciate a unique advantage of empiricism for the militant geographer: it assuages the ambivalence that will result for those who wish to "rescue" indigenous peoples with social science.[29]

The search for a quantitative measure of a scholar's worth has become a fetish. I know that the value of a body of research cannot be summarized by a number ("two articles"; "seven citations"). Nevertheless, because this contentious project is defended on the grounds that it is a serious academic research project, we have to ask: where is the research output? I think it is fair to argue that professional norms would expect that a team of scholars with ~$US 2.5 million in funding[30] would publish substantially more than a pair of mediocre, overlapping articles in a second-tier journal. Indeed I do not think it is too harsh to say that, *measured by disciplinary conventions and considering the investments of labor time and resources, the research results of the Bowman expedition to Oaxaca are only barely presentable as scholarly work.*

We should take one further step and hypothesize about the reason for the very existence of the *México Indígena* Bowman expedition. *The intellectual results of this project are weak not because of the intellectual shortcomings of those involved, but because of the involvement of the military in its creation.* A research project that is geared to "re-tool" social science methods to produce data for the military will be seriously hampered from producing original findings, for at least two reasons. The first is epistemological. The problematic and methodology of the research will be partly shaped by the military's interests. Consequently, the conceptual narrowing and empiricist character of the work (not to mention political or other restrictions on the funding) will invariably reduce the capacity of the scholars involved to critically and creatively question their premises and objects, to the great deficit of novel social thinking. The second reason is social. All labor is social, including intellectual labor; indeed, intellectual labor—particularly in the field of social analysis—is utterly dependent upon sharing and criticizing ideas openly. Yet with military involvement the character and implications of the research typically require secrecy as elements of the work are held back from open scholarly discourse (a point made sharply in O'Laughlan 2005).[31] Even if the research findings are made public, many other thinkers—those who oppose geopiracy, for instance—may be unwilling to engage those who collaborate with the

DOI: 10.1057/9781137301758

military on principled grounds. I have little doubt, for instance, that many human geographers would be unenthusiastic about spending their time reviewing research papers resulting from the Bowman expeditions. Such review is essential, not only for rigorously maintaining the standards of published research, but for sharpening ideas and arguments before publication. In short, social analysis requires open, public, fair-but-critical debate—to which military involvement is fatal.[32]

* * *

One of the important limits of such disciplinary critique is that it presupposes that I am in a position to positively discern the boundaries of the discipline of geography and its research norms. I cannot make such a claim. I include the proceeding remarks only because an immanent critique of the Bowman expeditions must still approach the question: what *are* they exactly? On one hand, they are clearly expressions of a more general disciplinary phenomena, i.e., geographical field research. On the other hand they are organized in a way that is distinctive. The synthesis of these points bring the conclusion that the Bowman expeditions reflect *US Army subcontracting of the analysis and mapping of human terrain to geographers.* To the extent that this synthesis is valid and coherent, then the Bowman expeditions are a species of the genus *military subcontracting to professional intellectuals,* another reflection of the "military-industrial-academic" complex (Cloud 2002). Its champions are the intellectual foot soldiers of David Petraeus and their funding comes from the US state/military, but their vision has been systematically supported by powerful civil institutions.[33] I have already mentioned the AGS, but public universities are implicated here too.

In 2004—*five years* before the first letters on the expeditions were written in Oaxaca—the University of Kansas and Fort Leavenworth signed a Memorandum of Understanding (MOU) specifying that the two institutions would "benefit by joining forces" (Wallace and Hemenway 2004, p. 1)—meaning, nota bene, that the scholars of the University of Kansas constitute a "force." To what end should these forces combine? To provide "leaders with a sound knowledge of the world."[34] With the leaders of the University of Kansas openly calling for military collaboration in Army-faculty "exchanges" and collaborative research (Ibid., p. 2), geographers at KU would only have needed to follow the prodding of their own university chancellor (a coauthor of the MOU) to align with the FMSO. The

DOI: 10.1057/9781137301758

MOU therefore signals that the conditions of possibility for the Bowman expeditions extend well beyond the discipline. The Bowman expeditions are pragmatic and derivative effects—marching orders, even.[35]

Contemplating the ostentatious and toady embrace of military funding by the leaders of this esteemed public university, while reflecting upon the imperious tone of the texts written in defense of the Bowman expeditions, called to my mind a verse of anti-imperial poetry. It is from Mark Twain's rendition of the Battle Hymn of the Republic, the version he "brought down to date" (1901,[36] p. 474) as the United States fatefully expanded its empire across Latin America, the Caribbean, and the Pacific:

> I have read his bandit gospel writ in burnished rows of steel:
>
> "As ye deal with my pretensions, so with you my wrath shall deal;
> Let the faithless son of Freedom crush the patriot with his heel;
> Lo, Greed is marching on!"

§4.5   The leaders of the Bowman expeditions, for their part, are not so openly avowed of military work. True, they have repeatedly stated that they wish to produce geographical knowledge in the very form that the US state/military needs it; they have gathered at Fort Leavenworth for strategic meetings; they express pride in their military funding (Herlihy 2010, at 35:30). Yet for all this, they insist that their concepts and methodology exist entirely independent of the US military. Indeed, in his 2010 AAG talk Herlihy even claimed that he had "invented" the participant mapping methodology de novo (2010 at 13:30) and that he was "among the earliest in Latin Americanist geography to recognize and to act on ethical responsibilities we have as researchers to indigenous societies" (9:35)—pretentious arguments to those familiar with this literature.[37] Even more remarkably, in reply to a question I asked about the involvement of the US military in shaping their research, Herlihy stated that "the FMSO had absolutely nothing to do with the development of our methodology" (1:22:00).[38] This elegant claim seems to allow the Bowman geographers a means to mentally separate the dollars from the data.[39] The military just gives them money; what they do with it, "participatory mapping," is something else entirely. Even if the paper trail did not point away from the veracity of this claim, it would make little sense. Would the US Army give away its money for no reason? If the authors of the Bowman expeditions gave a plausible explanation for the military's interest in paying for their work, this would make for a more effective alibi.

DOI: 10.1057/9781137301758

To conclude this chapter I recapitulate two points that follow from this analysis. The first concerns the place of the Bowman expeditions on the horizon of state/military involvement in geographical research. It is, of course, too soon to know; yet my hypotheses on this moment of militant empiricism are summarized in table 1, a schematic heuristic graph of Anglo-American geography's relations with the military. We are entering a new phase, one that reiterates the life and times of Isaiah Bowman beyond the eras of the Cold War, out of "geography's underworld," to using GIS to hunt for Bin Laden (Gillespie et al. 2009). In this new phase, field analysis returns to prominence—or rather, it "should" regain its prominence, Dobson would say—as an essential component of the "macroscope" (Dobson 2011) of global, GIS-integrated pattern analysis.[40] Fieldwork is essentially not only for securing detailed data about populations and resources in every corner of the world but also to train the would-be geographer qua counterinsurgency intelligence agent. Such fieldwork need not be directly executed by the military: indeed it is cheaper and cleaner for the military if the data is collected independently by geographers.

If I am right, it is only too appropriate that these practices unfold today in Isaiah Bowman's name. Allow me to quote again from Smith's magnificent biography:

> [C]iting the "growing influence of geography among military men," [Bowman] even urged the War Department [today's Department of Defense] to send some officers to the AGS [sponsor of today's Bowman expeditions] to advance their studies in "the field of geography as applied to military operations." He hedged about whether Latin American governments should be informed.
>
> What became of these plans tendering geography for the purpose of government spying is not clear. There have always been social scientists who have collaborated with government intelligence organizations, and from the time of the Roman geographer Strabo to the current CIA [headed by Petraeus], geography as a scholarly pursuit has traditionally operated as a handmaiden to the state. But the great majority of scholars have traditionally frowned on collusion with military intelligence operations, and scholarly associations often carry explicit prohibitions against spying. [...] What is remarkable about Bowman's injudicious peddling of geography and the services of the AGS is the lack of any sense that his eager cooperation with Military Intelligence, the government's premier spy agency of this period, in any way compromises his scientific integrity or endangers scientists. (pp. 89–90)

DOI: 10.1057/9781137301758

TABLE 1 *Anglo-American geographical thought and its relation to US militarism*

	Geography militant (Conrad)	Geography's underworld (Barnes)	Geography triumphant— yet divided	Geography counterinsurgent
Period (in US historical terms)	From British to US empire	US post war hegemony	US hegemony challenged	Declining US empire
Approximate dates	mid-1800s to World war II	WWII to ~1970	~1970 to ~2002	~2002 - present
Key geographical contribution	geopolitical analysis for empire	spatial analysis for Cold War: nuclear weapons	divided: social/critical theory, also GIS/spatial analysis	mapping human terrain
Exemplary figures	Isaiah Bowman	William Garrison; space cadets	David Harvey; Mei-Po Kwan	Jerome Dobson; Geoffrey Demerest
Key method	expeditions: fieldwork, mapping	analytical cartography & spatial modelling	close reading; GIS	expeditions redux, now organized for GIS integration

DOI: 10.1057/9781137301758

Smith's emphasis here on scientific integrity brings me to my second conclusion, concerning the implications of militant empiricism for what is usually characterized as "the health of the discipline," i.e., the possibility of continued political-economic stability for academic geography. The shift to geography counterinsurgent is coming about principally because of changes in US hegemony, not because of an objective "need" within the discipline (hence our disciplinary debate cannot be understood apart from broader political-economic changes—though it is not reducible to them). The "Oaxaca controversy" could have emerged in any number of "prototypical" regions mapped by the Bowman expeditions. It reveals that geographers stand awkwardly amidst a widening divide, one emerging within the discipline since the 1970s, with the (more or less concomitant) fluorescence of "critical" human geography on one hand and GIS on the other. The new round of debate over the heart of the discipline may well threaten the pax of the confident period that geography has recently enjoyed, with steady growth in the number of geographers trained and employed in the Anglophone world. *Our intuition of this threat explains geographers' anxious silence surrounding the "Oaxaca controversy."* Yet silence, as the saying goes, will not save us. Can we find the means and the concepts needed to think our way out of this disciplinary moment?

Notes

1   Kipp et al. (2006, no page) emphasize that the mapping and analyzing of human terrain "is based on unclassified or open-source information derived from the social sciences."

2   All citations from Conrad (1924): "fabulous geography" (p. 4); "circumstantially extravagant speculation" (p. 2); "adventurous action" (p. 2); "geography militant" (p. 6). Conrad has earned a poor reputation among geographers (see, e.g., Driver 2001) for these remarks. To be sure, his essay betrays a nostalgia for nineteenth-century expedition culture that is at best romantic, at worst masculine and colonial. Yet we should not be too quick to ignore Conrad's potential contributions to the postcolonial geographical imagination.

3   Bowman felt that US power would grow best through new global institutions charged with regulating capitalism on a global scale. "Whereas 'imperialistic leaders *demand* access to new resources or to increased

DOI: 10.1057/9781137301758

territory," [Bowman writes,] US expansionism was entirely justified because it
followed economic law" (Smith 2003, p. 188, italics in original).

4   Bowman (1918), cited in Crampton (2006, p. 733). Crampton's study details
how Bowman's intellectual leadership, and the work of the Inquiry more
generally, rested upon questionable presuppositions about the spatial fixity of
identity.

5   The authors of the Bowman expeditions deny that the name is significant,
apart from the fact that Bowman served as the director of the AGS from 1915
to 1935. On US imperialism in the post-9/11 era, see Harvey (2003), Gregory
(2004), Cockburn and St. Clair (2004), Grandin (2005).

6   In another paper (2006), Barnes elaborates upon this thesis, specifically in
relation to geographers who worked for the US Office of Strategic Services,
the forerunner to the CIA: "The discipline began to change in response to
its application, or in this case [the OSS] lack of application, to military ends.
The very experiences of some of the geographers at R&A [the Research
& Analysis branch of the OSS] as they tried to apply their geographical
training to war altered their conception of geographical research, helping
to propel the discipline to a different form [...]. It took a long time, but by
the late 1950s, parts of human geography began to model themselves on
research practices introduced at R&A emphasizing multi-disciplinarity,
team-based collaboration, problem-focused research, and rigour and
numerical methods. Approaches to war now shaped geographical thought"
(pp. 162–163).

7   One of the many evocative arguments Karatani makes in his most recent
book (2012) concerns the structure of historical repetition (pp. vii–xiii).
Karatani argues that although precise events do not repeat, the ancient
wisdom that "history repeats" may be true for a given structure, such as that
produced by the stages of global capitalism. In this view, the world finds itself
today repeating many qualities of late nineteenth century, viz: a declining
hegemonic state; imperialism; centrality of finance capital. I arrived at my
thesis about the return of geography militant in the guise of geography
counterinsurgent before reading Karatani's essay; I find that his analysis
provides a powerful, if necessarily schematic, basis to theorize this repetition.

8   Space does not permit an analysis of the ties between the emerging
importance of drone warfare, geopiracy, and militant empiricism, but see
Gregory (2011), Stanford Law School and NYU School of Law (2012) and
Shaw and Akhter (2012). Shaw and Akhter write (2012, p. 3): "The drone
dominates strategic US military thought and practice. In 2008, armed drones
flew over Iraq and Afghanistan for 135,000 hours [...] and dropped 187
missiles and bombs [...]. The US military plans to triple its inventory of
high-altitude armed and unarmed drones by 2020. In 2009 the US purchased
more unmanned than manned aircraft—and as General Petraeus, formerly

DOI: 10.1057/9781137301758

head of the US Central Command puts it, 'We can't get enough drones' […].
The military currently has close to 7000 unmanned aircraft, […] expected
to rise to 50 a day over the next 2 years and 65 a day by 2013." In this age,
all commanders are geographers of a sort: specialists in representing and
analyzing world-as-target. Boyce and Cash (forthcoming, no page) draw the
implications of the US military's use of drones for geography: "The routine
use of unmanned aerial drones by the U.S. military […] requires a growing
cadre of personnel whose task it is to review, process and share surveillance
data—generating additional logistical and bureaucratic hurdles for the useful
deployment and management of the technology and the information it
produces."

9   See Cockburn and St. Clair (2004); Gregory (2004).
10 I recognize that this is a generalization. Fortunately, a careful study of US
    counterinsurgency doctrine and its interrelations with social science is
    forthcoming from Oliver Belcher. Belcher's *précis* of this project—published
    in *Antipode* (2012) when he won that journal's 2011–2012 Graduate
    Student Scholarship—explains that the concept of "human terrain" and its
    "systematization" as HTS "serves a threefold purpose: to gather 'cultural
    intelligence' for commanders; to provide a supporting role for disrupting
    connections between local insurgencies and transnational terrorist
    networks; and to present a 'human face' to civilians caught in the web of
    combat operations" (p. 261). US counterinsurgency practices have long been
    practiced upon indigenous Latin America: see Grandin (2005).
11 The field manual is called "radical" by Sewall (2006, p. xxi). Apropos the
    novelty of the *Manual*: Price's (2009) article demonstrates that many of the
    ostensibly original claims and definitions in the *Manual* were plagiarized
    by military scribes. If there is anything new in the *Manual*, it stems from
    its emphasis on cultural-geographical analysis. The concept "human
    terrain" follows from this emphasis. Price (2009, p. 71) remarks on the
    distinctive quality of contemporary uses of scholarship to support military
    and intelligence needs: "That militaries commandeer food, wealth, and
    resources to serve the needs of war is a basic rule of warfare—as old as war
    itself. […] But requirements of modern warfare go far beyond the needs
    of funds and sustenance; military and intelligence agencies are evidently
    looking to commandeer scholarship in ways not intended by their
    authors." Unless, of course, the scholarly authors openly collaborate with
    the military.
12 On Petraeus' role, see Nagl's foreword (2007) to the *Manual* (2006).
13 Of course, one could argue that the United States did succeed in meeting
    certain goals in Iraq. For instance, economic policies were transformed
    along neoliberal lines and capitalist social relations entrenched through
    programs such as the US Army's "Operation Adam Smith" (see Gajilan

2004). On US attempts to make Iraq into a free market utopia, see Klein (2004).

14 Hayes was head football coach at Ohio State (1951–1978). His ridiculous slogan is taken seriously in Buckeye nation. The *Manual* uses a similar line: "the decisive battle is for the people's minds" (p. 49, §1-153).

15 Another way to put this is that counterinsurgency is always fundamentally *political*. This is the central claim in Edward Luttwak's (2007) visceral critique of Petraeus' counterinsurgency strategy. Luttwak notes that Petraeus and the other authors of the 2006 *Manual* "assume that it is simply an intelligence problem to identify the insurgents among the population. [...I]n fact it is a political problem, which always has a political solution, however unpalatable that may be" (p. 35). Luttwak further remarks that Petraeus et al. seem to believe "that a necessary if not sufficient condition of victory is to provide what the insurgents cannot: basic public services, physical reconstruction, the hope of economic development and social amelioration. The hidden assumption here is that there is only one kind of politics in this world, a politics in which popular support is important or even decisive, and that such support can be won by providing better government" (p. 34). Of course, there is not. While Luttwak does not specify the specific military-political strategy that he would endorse, he concludes his essay with the suggestion that the United States must either embrace the costs of long-term occupations or cease intervening abroad:

[T]he United States has preferred both in Vietnam long ago and now in Iraq to leave government to the locals. That decision reflects [...a] politics, manifest in the ambivalence of a United States government that is willing to fight wars, that is willing to start wars because of future threats, that is willing to conquer territory or even entire countries, and yet is unwilling to govern what it conquers, even for a few years. Consequently, for all the real talent manifest in the writing of [the 2006 *Manual*], its prescriptions are in the end of little or no use and amount to a kind of malpractice. All its best methods, all its clever tactics, *all the treasure and blood that the United States has been willing to expend, cannot overcome the crippling ambivalence of occupiers who refuse to govern*, and their principled and inevitable refusal to out-terrorize the insurgents, the necessary and sufficient condition of a tranquil occupation." (p. 42, my italics)

16 The geographers behind *México Indígena* discussed Iraq with Petraeus. They saw Iraq as an opportunity for another expedition. Thus the October 2007 Project Status Report refers to an unspecified presentation at Fort Leavenworth and a planned "Iraq Bowman Expedition," as well as a briefing to "the READIC Iraq team on the nature of property regime and global GIS place based research" (p. 1).

DOI: 10.1057/9781137301758

17 Does this breathless prose not betray a repressed homoeroticism? In fairness, there seem to be few Americans who are not enamored of this warrior hero. On June 30, 2011, the US Senate confirmed Petraeus as the new director of the CIA. The confirmation vote was 96 to 0: evidence that in the United States today there is no substantive political opposition to the riptide of ideas that carried Dobson and Herlihy deep into an ethical "grey area." If support for the Petraeus-inspired, "radical" approach to "counter-insurgency strategy" (see Sewall 2006) conforms to state power and provides status and research funding, and there is no organized opposition to it, in what direction should we expect geographical dialogue to flow?

18 I believe that their source is McFate and Jackson (2005). A similar definition appears in Kipp et al. (2006), who explain that the "human terrain" concept was developed through the FMSO at Fort Leavenworth in 2005–2006—i.e., the same time and place as the Bowman expeditions:

> To help address these shortcomings in cultural knowledge and capabilities, the Foreign Military Studies Office (FMSO) [...] at Fort Leavenworth, Kansas, is overseeing the creation of the human terrain system (HTS). This system is being specifically designed to address cultural awareness shortcomings at the operational and tactical levels by giving brigade commanders an organic capability to help understand and deal with "human terrain"—the social, ethnographic, cultural, economic, and political elements of the people among whom a force is operating. So that U.S. forces can operate more effectively in the human terrain in which insurgents live and function, HTS will provide deployed brigade commanders and their staffs direct social-science support in the form of ethnographic and social research, cultural information research, and social data analysis that can be employed as part of the military decision-making process.

19 I stress *may* because the same criticism could be made of *culture*, which is apparently a synonym for "human terrain"—yet "culture" has become the organizing object-concept of cultural anthropology.

20 See *México Indígena* Project Status Report, September 2006, p. 2. On the NGA, see note 21.

21 The argument that secure property rights lie at the basis of social order has a long and complex pedigree and it is beyond the scope of this work to review it. I note, however, that certain key texts in this tradition were written by empiricists, particularly John Locke.

22 See, e.g., Demarest (1995; 2003). Demarest is presently pursuing a PhD in the department of geography at the University of Kansas. I agree with activist-journalist Simón Sedillo that Demarest is the most compelling subject in this story—i.e., more impressive and more frightening than Herlihy or Dobson. Hence Sedillo, who deserves credit for "breaking" the *México Indígena* story to English readers (see Sedillo 2007, 2009), entitled his film on these matters

DOI: 10.1057/9781137301758

The Demarest Factor (Sedillo, 2010). The film's narrative suggests that while the geographers were wrong to enter into their fateful alliance with the US military, ultimately it is the military that bears responsibility for geopiracy. Notwithstanding my own criticisms of the film (e.g., its narrative fails to congeal around a sustained argument), it is to Sedillo's credit that he frames the problem in this way. Still, for scholars and especially geographers, the challenge remains to account for the conditions of possibility and effects of military collaboration.

23 This is not to suggest that the Mexican state uniformly opposes US military engagements in Mexican territory, which is certainly not the case. I do not doubt that it is the true that Bowman geographers "discussed our project at [...] political levels in Mexico and state authorities have extended significant interest and access to information related to politically sensitive research" (*México Indígena* Project Status Report for June–December 2007, p. 2). It is beyond the scope of this study to examine the role of the Mexican state in this story—a worthy topic that I hope someone else may take up.

24 "APPO" stands for the Assemblea Popular de los Pueblos de Oaxaca. This is the name given to assembly, first convened on June 17, 2006, that emerged to govern Oaxaca City during the political crisis. The APPO uprising—which has been known by other names—has generated a rich and voluminous literature, mainly in Spanish (e.g. Martínez 2008); for English texts see Denham and CASA (2008) and Gibler (2009), as well as the following works by geographers: Wright (2008), Mutersbaugh (2008), and Martin and Gática (2008). I reject the hypothesis, implied by some activists in Oaxaca, that the Bowman expeditions went to Oaxaca because of the APPO uprising. The timeline does not support this hypothesis. It is certain, however, that the Bowman expeditionists were concerned with the uprising on their work (see, e.g., their references to APPO in the *México Indígena* Project Status reports of January, June, and July 2007).

25 Those familiar with Foucault's (2008) lectures at the Collège de France of 1978–1979 will recognize in this passage the distinct features of a biopolitical approach to the social field. Given Foucault's interest in the police/military and biopolitics, it is worth remarking that Demarest has examined the "overlap between military and police in Latin America" (1995).

26 Measured by the standard metric, *JLAG* is not an influential journal. In 2011 the mean number of citations per article over the previous two years (aka the "journal impact factor") was only .21 (compare, e.g., with 1.5 for the *Annals of the AAG*, 1.4 for *Geoforum,* and 1.2 for *Antipode*) (data from http://www. scimagojr.com/journalsearch.php?q=29294&tip=sid). I recognize that impact factor is a problematic measure and, to be sure, the only way for one to evaluate the texts that result from the Bowman expeditions is to read them carefully.

27 These lines are almost repeated in the 2010 paper (p. 177). I am not suggesting that the authors are engaged in "self-plagiarism" (on which

DOI: 10.1057/9781137301758

see the AAG ethics statement) but rather "shingling," i.e., redundant publishing.

28 This might lead us to wonder about the potential results of the same research had it been conducted without state/military ties (the authors worked not only with the US military but also with the Mexican state, including PROCEDE itself). Yet without these ties the research would not have been the "same."

29 Herlihy et al. (2006), p. 11.

30 These funds were spent over a couple of years. My source here is Dobson's (2009) previous-cited statement that "we've received about $2,500,000." I suspect that this figure has grown considerably since 2009. A large sum in academic terms, this is peanuts in state/military calculus. The United States spent roughly one *trillion* dollars on the military in 2012 (see www.warresisters.org for details).

31 O'Laughlan warns that "the current political climate in the United States in the context of the 'war on terrorism' has frightened academic societies into censorship and self-censorship of research topics, results, and publication" (2006, p. 589). He concludes that while geographers are "entitled to engage in [...] military geography in a classified manner, [...] the journals of the Association [of American Geographers] should not be coopted in a manner that erodes their credibility and undermines the principles of transparency of academic work and publication" (p. 590).

32 For instance, on several occasions I have invited Professors Dobson and Herlihy to engage in an open, public debate on the question of whether geographers should accept funding from the US military. They have declined. Those who have seen Professors Dobson and Herlihy speak on their work recently know that, when this research is challenged in public, they tend to be unwilling to defend it in scholarly terms, electing instead to speak of it as *necessary* (both to geography and the US state/military), and/or to attack their critics via *ad hominum* argument. Whatever the ethics of such a stance, my point is that the lack of humility to question one's own research through open debate is fatal to thoughtful scholarship.

33 Remember the first letter from the Oaxacan communities called out "all the other institutions involved" (Hernández and Mendoza 2009).

34 The University of Kansas is not the only educational institution to support the Bowman expeditions. For instance, according to *México Indígena* Project Status Report Two, in July 2005 Dobson "met with the Rector of the UASLP [state university of San Luís Potosí, then the center of research for the Bowman expeditions in Mexico]" and found that the Rector was "very supportive of our project."

35 See also Bumiller (2012) on the return of US military officials to university classrooms.

DOI: 10.1057/9781137301758

36   The *Manual* (2006, p. 147) compares US Marine operations in Iraq
      (2004–2005) to those "used during the Philippine Insurrection (circa 1902),"
      the same that inspired these lines by Twain.

37   On participatory cartography, see chapter 1, note 2. Herlihy's earlier work in
      indigenous mapping has proven, to put it lightly, deeply controversial. For
      instance, Chapin and Threlkeld's study of indigenous landscapes (2001) is
      riddled with credible allegations about Herlihy's unprofessional practices
      with indigenous mapping projects in Honduras and Panama.

38   A minute later in his reply, Herlihy added that "a wide-eyed visionary,
      Geoffrey Demarest, together with Jerry Dobson, [. . .] brought this to the
      FMSO." The argument seems to be that Demarest and Dobson worked
      out the particulars in full, brought them to the FMSO, and found funding.
      Even putting aside the evidence that the Bowman expeditions geographers
      engaged in regular discussions with the US Army/FMSO regarding their
      methods and data—see their monthly reports—it seems strange for Herlihy
      to cite Demarest here since Demarest worked for FMSO.

39   Shortly after becoming Lord Chancellor to the British Crown, Frances
      Bacon, a man with considerable interest in "wealth, precedence, titles, [and]
      patronage" (Macaulay 1866, p. 316), "was charged with accepting bribes
      from people whose cases he had to judge. He admitted the charges, though
      claiming that no bribe or reward had ever actually influenced him" (Woolhouse
      1990, p. 9, my italics). Presumably the godfather of empiricism was only
      influenced by sense-experience.

40   Dobson (2011) proposes the term "macroscope" to name the spatial analysis
      of large patterns facilitated by the integration of fieldwork, satellite data, GIS,
      and so on. Like the microscope and telescope, Dobson sees the macroscope
      as a technology with potentially "revolutionary" implications for science.

DOI: 10.1057/9781137301758

5

From Geopiracy to Planetarity

Abstract: *Gayatri Chakravorty Spivak as a geographical thinker ~ Her proposal that the planet overwrite the globe ~ planetarity ~ Qadri Ismail's critique of empiricism ~ His conception of abiding ~ Putting these concepts to work ~ The question of fieldwork ~ A text by Kiado Cruz ~ The Xidza Declaration.*

Keywords: abiding; Kiado Cruz; fieldwork; geographical objects; Qadri Ismail; planetarity; postcolonial theory; postempiricism; Gayatri Spivak; *Xidza* declaration

Wainwright, Joel. *Geopiracy: Oaxaca, Militant Empiricism, and Geographical Thought*. New York: Palgrave Macmillan, 2013. DOI: 10.1057/9781137301758.

> Enlightenment, understood in the widest sense as the advance of thought, has always aimed at liberating human beings from fear and installing them as masters. Yet the wholly enlightened earth is radiant with triumphant calamity.
>
> Horkheimer and Adorno (1947), p. 1

§5.1 Many human geographers have become familiar with postcolonial theory, particularly through the writings of Edward Said. Said's *Orientalism* (1978) inspired numerous examinations of geographical representations of former colonial societies.[1] Notwithstanding Said's influence and the emergence of a literature on "postcolonial geography," however, the discipline's engagement with postcolonial theory has been rather modest.[2] Here I argue that the work of two particular postcolonial critics, namely Gayatri Chakravorty Spivak and Qadri Ismail, are of special significance for those geographical thinkers who wish to confront militant empiricism.[3]

The best place to begin is still Spivak's enigmatic, brilliant essay, "Can the subaltern speak?" First drafted in 1983, it was initially published in an edited collection in 1988; Spivak revised its argument for the "History" chapter of her *Critique of postcolonial reason* (1999). If it is difficult to crystallize the meaning of this essay for geographical thought, it is partly because these multiple versions circulate and are read in wildly different ways. Yet we should not overstate the difficulty in grasping the essay's problematic, since it is announced in the abstract's very first sentence: "An understanding of contemporary relations of power, and of the Western intellectual's role within them, requires an examination of the intersection of a theory of representation and the political economy of global capitalism" (Spivak 2010, p. 237). The abstract explains, therefore, that the essay will examine the relations between power and the production of knowledge by Western intellectuals through an analysis of global capitalism and the theory of representation. And indeed, the essay scrutinizes the place of Western intellectuals in the operation of power, specifically by problematizing both the world (or *geo-*) and the problem of representation (or *graphē*) that, we learn in the essay, present the Western intellectual (e.g., a geographer) with certain impassible challenges, of aporias. Now there is much more to say about this essay, but I think we need go no further than to see its relevance for geographical thought. Sanders notes that Spivak's style of thought "may be viewed as

DOI: 10.1057/9781137301758

an *experiment in reading the world*" (2006, p. 32, my italics). This elegantly summarizes why Spivak's thought offers a powerful corrective to militant empiricism—particularly if we emphasize the deconstructive quality of "reading" here (on which see Derrida 1967, translated by Spivak).

Nevertheless, Spivak is not widely regarded as a theorist of geographical thought. Her work is seldom taught in geography seminars and few geographers have written thoughtful commentaries on her work.[4] This is unfortunate. To be fair the spatial and geographical elements of her thought have not been thematically organized around the conventions of our discipline. That is to say, Spivak has not been disciplined as a geographer—and for this we should all be thankful. This leaves geographers with the homework of reading Spivak's oeuvre, where we cannot help but find a set of concepts—worldliness, Eurocentrism, spacing, planetarity, critical regionalism, fieldwork, territorial imperialism, "other Asias," and so on—that cry out for our careful attention. To run the risk of simplification, Spivak's postcolonialism has *always* exhibited a radical critique of the West's validation of expeditionary experience as a basis for representing the world.[5] Spivak's work thus braces the feminist critique of the masculine bravado of geography's muddy boots.[6] Remember that Dobson's dream is one of total command of world via a "macroscope" (2011). It is easy to poke fun at a man's desire for an enormous phallus, but the problem is not Dobson but militant empiricism. How might we rearrange this disciplinary desire?

An answer of sorts—an impossible answer, perhaps, but one nonetheless—is introduced by Spivak in the third chapter of *Death of a discipline* (2003) where she argues for seeing comparative literature and area studies—and I think we should add geography to her list—as "planetary rather than continental, global, or worldly" (p. 72). Here and in subsequent writings Spivak sketches a concept that may be fruitful for those geographers who wish to confront geopiracy: planetarity. In its earliest formulation, this concept came in the form of a proposal: "I propose the planet to overwrite the globe" (2003, p. 72). Spivak elaborates:

> Globalization is the imposition of the same system of exchange everywhere. In the gridwork of electronic capital, we achieve that abstract ball covered in latitudes and longitudes, cut by virtual lines, once the equator and the tropics and so on, now drawn by the requirements of Geographical Information Systems. [...] The globe is on our computers. No one lives there. It [the globe] allows us to think that we can aim to control it. (2003, p. 72)[7]

DOI: 10.1057/9781137301758

It may seem that Spivak is proposing that we exchange concepts; Let us no longer speak of globalization, the global scale, and the like; instead let us think of ourselves living on a planet. This would present a solid starting point, though Spivak hastens to add that planetarity is not "amendable to a neat contrast with the global" (Ibid.). "The planet," she explains, "is in the species of alterity, belonging to another system" (2012, p. 338), one beyond our control and even representation.

As I follow Spivak, her conception of the world qua "planet" differs from the common conception of the world qua "globe" in that planet is one of those things that can never be a thing, but a thing-in-itself[8]: something that we know is there, though we can never directly grasp as an object with our senses. (Hence Spivak's use of "planet" is emphatically not to be interpreted in the strictly empirical sense as an object called "Earth," the third planet from the sun.) And yet what makes "planetarity" even stranger is that the "planet" invoked by the term is the world we inhabit—indeed, we *are* it (Spivak 2012, p. 338).[9] By contrast, the "global" and "globalization" can be concepts only for objectifying *what is*—so that we locate, map, measure, calculate, target, and so on. Global-talk is given to representations of the world in the mode of empiricist geography. Planetarity reflects our being in a singular and peculiar sense, viz: our being responsible to our being incapable of representing ourselves.

But how, if not as empiricists, are we to think planetarity? Spivak answers: "'planet' is, here, as perhaps always, a *catachresis for inscribing collective responsibility as right*. Its alterity, determining experience, is mysterious and discontinuous—an experience of the impossible" (2012, p. 341, my italics).[10] Here Spivak is attempting to think our worldliness through a conception of being-with, an ethics, that we cannot ever ultimately *achieve*; hence, an aporetical ethics: a guide to practical being and doing *as if* one could be ethical toward the other, all the while knowing that this is impossible. In this sense Spivak's proposal to overwrite globe-talk with planetarity is not only a critique of geographical empiricism (though it certainly is this) but also a reminder of empiricism's provenance and a call to an impossible encounter to come.[11] Planetarity displaces experience from its empiricist basis (where it remains unthought), since our experience of being in the world cannot account for the "mysterious and discontinuous" experiences—birth, love, death, and so on—that entangle the planet (and remember, that is us) through the double bind of being responsible.[12]

Such talk of aporias may well leave many geographers scratching their heads and wondering whether there is not another way to put this—a

DOI: 10.1057/9781137301758

way that seems less impossible. This is where I think Qadri Ismail's extension of Spivak's project proves to be essential. Not only because it makes these problems much clearer—brutally clear, for the empiricist—but also because Ismail takes up a question that I think Spivak's work has frequently alluded to, but not thematized: How are we to rethink the question of representing space and place after the postcolonial critique?

Allow me to answer this question by briefly recapitulating the introduction of Ismail's remarkable book *Abiding by Sri Lanka* (2005). This study begins with an apparently simple problem. Meditating upon the challenge presented by a particular reading of the "post-July 1983 conjuncture" (p. xii), Ismail asks: how is it possible to read events as taking place *in Sri Lanka*, i.e., specific to a reality *there* and *then*, not *here* and *now*.[13] How, except through such distancing and empirical reduction, are we to comprehend Sri Lanka? Are we not always already committed to grasping events as being placed in such places, within an implicit national frame, and contrapositively of thinking all places as the result of historical events that happened? If so, does that not foreclose the possibility of responsibly thinking with other spaces? These questions are definitive for geography's disciplinary formation, and like Spivak, it is our good fortune that Ismail poses it anew and in a way that is not marked by his having been disciplined as a geographer. He argues that our "current epistemological or *disciplinary* moment—which I prefer to characterize as postcolonial and postempiricist" (p. xiii)—requires that we reaffirm and consolidate the advances made by postcolonial theory.

How, you may ask, might the "consolidation" of postcolonial thinking change how we conceptualize the world? *Abiding by Sri Lanka* (hereafter *Abiding*) does not pose this question as a problem for geography per se, but of two of our cognate disciplines: anthropology and history.[14] This could lead geographers to conclude that *Abiding* is not a critique of geography, or has little to say to geographers. I think this conclusion would be errant. What makes, by my reading, *Abiding* so important for geographical thought is the way that it rigorously problematizes the question of representing place and space. Insofar as this question remains fundamental to the work that is disciplined *as* geography, *Abiding* has much to teach. Its critique goes well beyond anthropology or history (not to mention Sri Lanka) and concerns empiricism generally, more specifically the application of empiricism to social life—the disciplinary matrix of the social sciences. Hence we read in *Abiding*'s introduction

DOI: 10.1057/9781137301758

that the crucial task of postcolonial critique today is to "make the social sciences [...] account for their complicity in naturalizing [an] empiricist [...] understanding of the social" (p. xiii). What makes the empiricist approach to the social our problem? Ismail elaborates, proposing a neologism, "postempiricism," to refer to his postcolonial anti-disciplinary critique:

> By deploying [postempiricism], I want to signal that the break to be made—which cannot be a complete or clean one, since the "post" signifies that one is still quarreling with, trying to displace, and therefore complicit with, empiricism—is not so much with the concept of structure but empiricism, and its postulate, the empirical. As Jacques Derrida reminds us, in his critique of Lévi-Strauss,[15] structuralism promised such a break but did not deliver. Empiricism, as [Derrida] put it there, is the "matrix of all faults" infecting the social sciences. So postempiricism refers, very broadly, to those literary critical persuasions that begin from this position, to those that take reading and/or textuality [...] as their point of departure. (p. xiv)[16]

Postcolonialism then would be one of those "literary critical persuasions" that enact the radical break that poststructuralism failed to deliver. To what end?

Let us consider again the problem of grasping places, regions, or territories—the *Rincón*, the state of Oaxaca, or Mexico, for instance—as geographical objects. As we have seen, the Bowman geographers were able to proceed expeditiously to this place, conduct fieldwork, collect data, and capture this object in a GIS for the US Army. In all of this there would never have been any equivocation about the need for empirical facts, nor concern for reading—only a desire to have a better tool, a bigger "macroscope" (Dobson 2011) with which to grasp and command the world. In Ismail's terms, for militant empiricism the geographical object "is conceived as transparent, outside language and the process or play of signification" (2005, p. xiv). By contrast, a postempiricist reader "does not conceive of herself as an autonomous or agential subject, conceives of her object as also subject, as simultaneously subject and object" (2005, p. xiv). The point is to work toward the transcendence or destruction of the very distinction between subject and object.[17]

On this point, Ismail's critique extends from a long genealogy. Perhaps the most pertinent illustration is Derrida's critique of Lévi-Strauss (cited by Ismail), but we should also include Kant's (1787) critique of Hume's empiricism (not discussed by Ismail).[18] To appreciate the relationship between postcolonial geography and the latter it is useful to turn briefly

DOI: 10.1057/9781137301758

to Vinay Gidwani's brilliant study, *Capital, interrupted* (2008, p. 22). Reading a British colonial text from 1894 on "waste lands" in Bengal, Gidwani writes:

> The impressionistic remarks [in the 1894 text] were able to carry their degree of conviction, I suggest, precisely because they were generated by a network of premises that had already rendered "India" as an object in imagination. [...] In short, colonial empiricism exemplified Kant's critique of Hume, in that sense data—and statements that claim their basis in "brute facts"—are never free of presuppositions.

By Ismail's rendering, this peculiar quality of colonial empiricism has been generalized in and through the very structure of the social sciences. Postempiricism, then, would not comprise a puritanical effort to produce truth without any trace of the empirical, but rather rooting out the "network of premises," in Gidwani's terms, that allow us to grasp geographical objects in the empirical, or imperial, fashion. To what end?

Ismail's immediate aim—"to produce a different object [...] for the postempiricist and the postcolonial" (p. xv)—may seem obscure. As he explains, the task at hand is to understand places "not geographically, or through its ally, area studies, but as a *debate*; not as an object that exists empirically but as a *text*, or a group of texts" (p. xvii, my italics). This is not to suggest, of course, that there is no world or that places do not exist except textually. Rather it is to argue that we cannot access the so-called real world except as it is mediated textually, or through language and concepts, and that if we wish to abide by any place we have a responsibility to take account of such mediation. But how, concretely speaking, would we abide by a place when grasped as "a debate" or "a group of texts"? Why, or where, is there abiding? Ismail's reply:

> To abide by a place [...] cannot be to physically reside in it. One cannot, after all, physically reside in a text. Rather, it means to display a commitment to attending to its concerns, to intervening within its debates, to taking a stand—to sticking one's neck out if necessary [...]. It means to display patience, to stay with it, endure it, work with it, even if it appears [...] unbearable, unending, unendurable. (p. xxx [30])

To summarize formulaically: abiding emerges not through research but from responsibility. Abiding exists because of *thinking being worldly*.

Such an aporetical conclusion repeats, if not the language, at least the tone of Spivak's discussion of planetarity. The parallels go beyond this of course. Among many other shared aspects, neither Ismail's nor Spivak's

DOI: 10.1057/9781137301758

concepts can escape their own genealogies (as they would be the first to admit). *Abiding* never accounts, by my reading, for the Christian metaphysics laden in its central metaphor.[19] Likewise Spivak's planetarity will be misunderstood as an appeal to our being planetary citizens, in the very spirit of the liberal cosmopolitan-environmentalism that she seeks to displace. She anticipates such reading and tries to turn it away:

> If we think dogmatically (to borrow Immanuel Kant's phrasing in English translation) of planetarity as contained under another concept of the object which constitutes a principle of reason and then determine it in conformity with this, we come up with contemporary planet-talk by way of environmentalism, referring, usually though not invariably, to an undivided "natural" space rather than a differentiated political space. This smoothly "translates" into the interest of globalization in the mode of the abstract as such. This is the planet as an alternate description of the globe, susceptible to nation-state geopolitics. It can accommodate the good policy of saving the resources of the planet. My use of "planetarity," on the other hand, does not refer to an applicable methodology. (Spivak 2011)[20]

Planetarity, to repeat, is not a methodological concept. Neither is abiding. Not tools that one uses to do geography, they are contributions toward geographical thought. We may appreciate them as contributions to this end only if such thought is enacted through a setting to work of worldly responsibility (which cannot be realized qua social science). This is why the *polemos* with militant empiricism will not be "won" by postcolonial theory. It is a question of learning to do something else.

* * *

But *where* should we do this "something else"? Shall we go to the field? Shall we conduct "fieldwork"? What, if anything, might constitute a form of fieldwork which is not "expeditionary"? Where could one find a field for abiding? Let us briefly consider these questions.

To begin it will be useful to have some definition of fieldwork. Consider that provided by a well-known geographer in the Latin American, Sauerian tradition, Barney Nietschmann: "Field research means leaving the university, the library, and the laboratory to go somewhere—near or far—to obtain firsthand information from firsthand investigations" (2001, p. 176). Nietschmann's pragmatic definition betrays an empiricist epistemology.[21] (The repetition of "firsthand," e.g., in his definition is code for sense-experience, empiricism's keyword.) And in his conception, the

DOI: 10.1057/9781137301758

defining quality of fieldwork is the act of *distancing*—"leaving the university" and going "somewhere," anywhere—to *obtain data*. It is a question of spacing: of creating a space between the *field* (data source/periphery) and the *university* (center of calculation/center).

It is safe to say that Ismail, at least, would encourage geographers to cease doing fieldwork in Nietschmann's sense. Abiding implies the destruction of the "here"/"there" structure of empiricist fieldwork, where the model is to go from the "here" of the University to the "there" of the field (however "near or far" the "there" may be, it is always a *somewhere-other*). Rather, abiding requires us to work toward the destruction of such an opposition:

> Fieldwork, however long in duration, attending to native voices, however far they may be from the capital or dangerous the enterprise, is not central to the question of abiding by. Neither is subjectivity. The pivotal factor, quite simply, is whether the *text* addresses, feels itself accountable and responsible to, the questions and concerns *only* of the powerful epistemological space of the West or to those of Sri Lanka as well. (Ismail 2005, p. xxxi)

I find this formation rigorous and in keeping with my conception of the postcolonial critique of geography. Yet I am also aware that dislocation, as it is enabled by international travel, can often facilitate the very rearrangement of desire that might lead one to question the privileged epistemological space of the West.[22] We have, then, a double bind. If there is a way out of it, and I am not sure that there is, it would require practically living in such a way that we may unlearn disciplinary habits.[23]

To this end Spivak has proposed in a number of texts and interviews to think fieldwork as a kind of practiced "hanging out" (e.g., "I became involved in hanging out in that subaltern space, attempting [...] to think it a normal teaching scene": 2010, p. 230). As I read her, Spivak would endorse Ismail's visceral critique of fieldwork, only to a point, for her texts also enable a catachrestic reading of "fieldwork," where fieldwork may be understood not as the usual empiricist spacing (again, the distancing of the here/there so that the empirics of the "there" can be assembled within the epistemological space of the West). What does fieldwork become in her catachrestic reading? On one hand: a playful being-there, an encountering, best left untranscoded; on the other: an opportunity to "suspend previous training in order to train yourself" (2005, p. 524).[24]

This is not a formula that graduate students will rush to cite in their research proposals to funding agencies like the NSF and SSRC. Neither

DOI: 10.1057/9781137301758

is Spivak's conception of fieldwork *useful* for geographers. This is good news, at least for those of us who share the conviction that postcolonialism and postempiricism still have a lot of work left to do (Ismail 2005, p. xvi). If there is bad news here, and I am not sure there is, it is that one cannot ever know whether one is properly doing fieldwork in Spivak's sense. We cannot escape the double bind or the responsibility it engenders. Yet recognizing one's complicities and responsibilities—to thinking and abiding—is, in itself, an important accomplishment. Jerry Dobson is not wrong that the "'War on Terror' requires a [...] commitment to geographic fieldwork" (cited in Herlihy et al. 2006, p. 5). In the face of such requirements we should be open to the possibility that some apparently useless concepts are more worthy of our thought than is answering the call of counterinsurgency.

The thought of planetarity is not only useless. It reflects an almost violent epistemic encounter with the singular and incalculable, a secular confrontation with the random accident of one's own being in the world. Allow me to cite the eloquent words of Najeeb Jan (2011):

> What kind of thinking can redirect us to the question of what it means to be in this world? This is a kind of thought that does not submit thinking to the exclusive rule of the distinction between exchange and utility. This is not to insist upon an abstract thinking for its own sake, but rather the valuing of forms of thinking and questioning that cannot immediately be harnessed within a calculus of direct practical advantage. This form of thinking exhorts us to uncover, contest and transcend the reified sediments of our deepest metaphysical assumptions, and hence possibly our most deeply cherished assumptions about [...] the nature of the self and world.

If there is a practical or straightforward "lesson" for geographers from this line of thinking, it concerns something that is arguably fundamental to the discipline, or at least the concept of fieldwork that legitimates empiricist geography: the here/there distinction. The lesson is that we should not posit a place or region—e.g., the *Rincón* or *Oaxaca*—as the "field," the "there" from where we derive data. And any discussion "here"—e.g., in the AAG or the US academy—should be responsibly unsettled by the critique of the here/there distinction. Asking or expecting Oaxaca to politely enter into our quiet and tightly circumscribed ethical discussion is not enough. We have yet to open a space where "Oaxaca" no longer serves as the silent backdrop of "our" controversy. To do so would require abiding by Oaxaca. Fieldwork, catachrestically understood, may or may not be useful to this end.[25]

DOI: 10.1057/9781137301758

§5.2   The two letters written in early 2009 are by no means the only texts from Oaxaca that deserve the attention of geographers. On the contrary, the discourse within Oaxaca, and Mexico more generally—for this controversy has received national attention—has taken the form of news articles, essays, poems, and at least one film. It is beyond my capacity to provide an archaeological reading of this discourse here.[26] Instead I offer brief remarks on two especially significant texts written in Oaxaca since 2010.[27]

The first is an essay by Kiado Cruz written in 2010 and published with the collection on the controversy in *Political geography* (I discussed Byran and Agnew's contributions in chapter 3). Cruz, a young indigenous activist from the village of Yagavila—one of the communities in the *Rincón* where the Bowman expeditions attempted to conduct their research, but were essentially expelled[28]—frames our controversy against a remarkable panorama of calculation and expropriation:

> Today we stand before a process of worldwide reorganization, where land is first measured and then alienated, where its resources are first documented and then appropriated, to be used for a new cycle of investment and accumulation. Facing this situation, it is important to ask: what do these new processes of accumulation offer us and what good, if any, can come of them? What progress and development have we received from them? What are the cultural, social, technological and economic benefits for our people? These are some of the many questions we can ask in an attempt to understand the interests behind the *México Indígena*: Bowman Expeditions project that came to work in our community of Yagavila in 2006. (Cruz 2010, p. 420)

The Bowman expeditions only emerge in this text here, after this intensely geographical description of the broader horizon of the dispute. With these questions—which Cruz asks but does not answer—the essay's preamble closes and Cruz summarizes the local facts:

> The researchers and students led by Peter Herlihy came before the General Assembly of our community in August 2006, claiming that the objective of *México Indígena* was to conduct participatory mapping[.... M]any people asked specifically about the project's financing, to which the researchers replied "The financing of the AGS Bowman Expeditions can come from any source, public or private." It was not explicitly stated that financing came from the Foreign Military Studies Office (FMSO) and we were not informed that the data they obtained would be given to the FMSO of the U.S. Army. (Ibid.)

Roughly two years after the exchanges described here, Cruz and other indigenous activists learned fuller details about the Kansas-Army

DOI: 10.1057/9781137301758

DECLARACIÓN XIDZA SOBRE GEOPIRATERIA

Los que suscribimos, Autoridades Municipales y Comisariados de Bienes Comunales de las comunidades de San Juan Tepanzacoalco, Santa María Zoogochi, Santa Cruz Yagavila, Santiago Teotlaxco y San Juan Yagila, reunidos en la comunidad de San Juan Yagila el día 24 de julio de 2011 en la sala de reuniones de la Agencia Municipal, después de haber reflexionado sobre lo sucedido en las comunidades de San Juan Yagila y San Miguel Tiltepec en el año 2006, cuando se realizó en estas comunidades el proyecto México Indígena, que formó parte del Proyecto Global denominado Expediciones Bowman, impulsadas por la Sociedad Geográfica Americana (AGS por sus siglas en Inglés) y la Oficina de Estudios Militares para el Extranjero (FMSO por sus siglas en inglés) perteneciente al Ejército de los Estados Unidos, declaramos lo siguiente:

- No estamos de acuerdo con la forma en que se realizaron los estudios geográficos en las comunidades de San Juan Yagila y San Miguel Tiltepec por parte del equipo del Proyecto México Indígena, entre los años de 2006 y 2008, porque no se informó a estas comunidades del origen de los recursos que se utilizaron para la realización de esta investigación, ocultando expresamente la participación del Ejército de los Estados Unidos, violando de esta manera el derecho al consentimiento libre previo e informado que las comunidades indígenas tenemos reconocido en la Declaración de las Naciones Unidas sobre los Derechos de los Pueblos Indígenas; asimismo respaldamos a ambas comunidades en los problemas que puedan tener en lo posterior a raíz de las investigaciones realizadas

- Nos pronunciamos por hacer un solo frente organizado entre las comunidades de nuestra región conocida como el Rincón de Ixtlán.

- Buscaremos hacernos llegar la información necesaria sobre los pros y los contras de los proyectos y programas gubernamentales y no gubernamentales que se ofrezcan a nuestras comunidades, a fin de que antes de decidir si aceptarlos o no, se aplique el principio del consentimiento libre, previo e informado.

DOI: 10.1057/9781137301758

- Demandamos que se pague a nuestras comunidades de manera incondicional y compensatoria, recursos económicos suficientes por la conservación de los bosques que existen en ellas ya que está demostrado que son las comunidades indígenas las que han conservado los bosques y selvas de México, así mismo que esto se haga con recursos públicos, para no caer en manos de empresas transnacionales que solo están interesadas en lucrar con nuestros bienes y lavar su culpa por la crisis climática que han provocado en el planeta.

FIGURE 3   *Declaracíon Xidza Sobre Geopiratería*

DOI: 10.1057/9781137301758

connections. They brought word to the communities, leading to the two letters in early 2009.

I emphasize that Cruz's text matters—like the earlier letters from Oaxaca—not merely for the specific details they provide concerning the Bowman expedition to Oaxaca, but also for its explicit postcolonial insights. Consider, for instance, Cruz's claim that "we have always seen a strong link between geography and the interests of the military industrial complex, especially in recent attempts to create worldwide property databases" (Ibid.). Throughout his essay Cruz asks us to imagine a way of thinking geographically that is not bound by this "strong link." He emphasizes the ontological dimension of this task, asking how we might "oppose the usurpation—conducted through the use of new devices and electronic systems—of the communal goods which are most intimate and important to our being." This is a critique of a form of geographical thought of which the Bowman expeditions are only one instance. Cruz's text says little about the Bowman expeditions; it appears motivated not exclusively by the controversy but also by a more general and dangerous "process of worldwide reorganization" (Ibid.). The text's authority derives not only from its conditions of production or authorship (a would-be research subject writing back to *el norte*) but also from its status as a critique of the very ordering of the world (the "worldwide reorganization" that frames the essay). While the essay delivers a critique of the Bowman expeditions, it also offers—unlike, e.g., Bryan's essay in the same issue of *Political geography*—a non-empiricist geographical depiction of the *Rincón*, born of *polemos*. One that deserves a response.

So too a second recent text from Oaxaca, entitled the *Declaración Xidza sobre geopiratería* (hereafter *Xidza* declaration). The *Xidza* declaration was written, stamped and signed by the Municipal Authorities of four rural Zapotec communities after a day-long discussion in Yagavila, Oaxaca, on July 24, 2011 (see figure 3).[29] The declaration begins:

> [H]aving reflected upon what happened in the communities of San Juan Yagila and San Miguel Tiltepec in 2006 when the *México Indígena* project was carried out [by] the Bowman Expeditions, promoted by the American Geographical Society and the Foreign Military Studies Office [of the US] Army, [we] state the following:

There follows a series of points. The first reads:

> We do not agree with the manner in which the geographical studies were carried out in the communities of San Juan Yagila and San Miguel Tiltepec

DOI: 10.1057/9781137301758

by the *México Indígena* project team between the years of 2006 and 2008, because they did not inform these communities as to the origins of the resources they used to carry out this research, specifically concealing the participation of the US Army, and in this way violating the right to free, previous and informed consent which for us as indigenous communities is recognized in the [UN] Declaration on the Rights of Indigenous Peoples. (Municipal Authorities 2011, p. 1)

I note in passing that all three elements of the discourse on the Bowman expeditions (discussed in chapter 3) resonate here. The *Xidza* declaration elaborates upon these three themes and the injustice of geopiracy before concluding with a comment upon something that may seem, to impatient geographers, as a wholly separate issue: climate change.

> The forests and jungles of Mexico, the same that comprise for us public resources, should not fall into the hands of transnational corporations which are only interested in exploiting our commons and hiding their culpability for the climate crisis that they have brought upon the planet.

Can we not read this text as a provocation to planetarity? As a demand that we think otherwise of our responsibility to the commons?

These questions deserve careful elaboration but I conclude instead with one final remark on the *Xidza* declaration. Its existence—including not only the words on the page but its stamps and signatures—signals that as of July 24, 2011, all of the communities in Oaxaca that were studied by the Bowman expedition *México Indígena* have formally and publicly condemned the project, called it "geopiracy," and demanded justice.

How shall we respond?

Notes

1   On *Orientalism* and geography, compare Gregory (2004) and Wainwright (2005).

2   I recognize that this is a generalization and that I may be missing some good work. For exceptions consider the writings of Kiran Asher, Sharad Chari, Vinay Gidwani, Najeeb Jan, and Tariq Jazeel, among others. For my take on postcolonial geography, see Wainwright (2008).

3   As Spivak has come to be seen as *the* postcolonial critic, it is confusing that she resists the label as an adjective for her work—in fact she has described her project (in 1999) as a specifically *feminist critique* of postcolonial reason. By contrast, Ismail uses "postcolonial" affirmatively and claims that it

DOI: 10.1057/9781137301758

should be "*consolidated,* affirmed, abided by" (2005, p. xiii). Their work is not exclusively "postcolonial." Their project is influenced markedly by Marx, Gramsci, Fanon, Althusser, Derrida, Subaltern studies, and Chatterjee, not to mention feminism. So if you prefer you could call this "postcolonial Marxist feminist critique" (a formula often used to describe Spivak).

4   A recent exception—where Spivak is read as an immanently geographical thinker—is Jazeel (2011), a paper singled out for praise in Spivak's 2012 *Antipode* lecture.

5   For instance, it follows from Spivak's teachings that when "the East"—or Oaxaca for that matter—can only be "imagined as an *area* of study and the global South is objectified," as is necessarily the case with militant empiricism, her sort of "deconstruction can encourage a response in ethical singularity by attending to idiomaticity of language and codes of cultural production" (Sanders 2006, pp. 48–49, my italics). Remember that the Bowman Expeditions are named for a man who said that the essence of Geography is that it "invites exploration [. . .] emphasizes location [. . .] involves measurement": the Eurocentrism of such a conception of geographical thought is palpable.

6   The feminist-geographical literature on fieldwork is extensive and rich. See especially Katz (1994) and Sundberg (2003); for a synoptic review, Sharp and Dowler (2011); for a historical study of the AGS vis-à-vis the masculine-fieldwork tradition, see Morin (2011); on alternative conceptions of feminist transnational fieldwork, in addition to those just cited, see Nagar (2002) and Swarr and Nagar (2010).

7   This passage is cited and discussed in Moore and Rivera (2011), p. 27; compare Spivak (2003) and (2012). As I read it, Spivak's critique of the global reiterates Heidegger's critique of modernity (1938). This is an "interested connection."

8   On thing-in-itself see Kant (1787); also Karatani (2003, part I).

9   This is my rephrasing of her line. In its first formulation (2003, p. 72) this sentence concludes: "yet we inhabit it, on loan." I suspect that Spivak decided to change "on loan" to "are it" because the former implies a sort of theological economy while the latter emphasizes practical/ontological unity. In either phrasing there remains an echo of Heidegger's claim that "in accordance with the kind of being belonging to it, Dasein tends to understand its own being [Sein] in terms of *the* being [Seienden] to which it is essentially, continually, and most closely related—the 'world'" (1927, p H15).

10 Spivak's argument is, among other things, a critical reading of cosmopolitanism (see Jazeel 2011).

11 Hence when Spivak "invoke[s] the planet, I think of the effort required to figure the (im)possibility of this underived intuition" (2003, p. 72).

DOI: 10.1057/9781137301758

12   On the double bind, see Bateson (1969) and Spivak (2012).

13   July 1983 is usually given as the starting point of the military conflict between the Sri Lankan state and the Tamil militants. As always, the story is more complex; see Ismail (2005). All quotations in this and the subsequent paragraph are from Ismail (2005).

14   Its critiques of these two disciplines are rigorous and absolute (contestable too, but for me persuasive). By contrast, literariness and the capacity to abide with places are treated positively.

15   The reference is to Derrida (1966); see also Derrida (1967).

16   By my reading of Spivak, textuality (or social textuality) is a deconstructive lever from Heidegger's "worlding." Thus she explains that textuality is closely "related to the notion of the worlding of a world on a supposedly uninscribed territory. When I say this, I am thinking basically about the imperialist project which had to assume that the earth that it territorialised was in fact previously uninscribed" (1990, p. 1).

17   This passage concludes: "Most important, she works toward the demise of this opposition [between subject and object]."

18   See Preface, note 6.

19   The condemnation of John 5:38 rhymes with Ismail's critique of social science: "ye have not his word abiding in you" (American Standard Version; see also John 1:33 and 14:25). So too the first stanza of the hymn, "Abide by me," concludes with two lines that characterize *Abiding*'s impossible task: "When other helpers fail and comforts flee | Help of the helpless, O abide with me."

20   I thank Professor Spivak for discussing these questions with me and for sharing access to this text. See also *Planetary loves* (Moore and Rivera 2011, especially Spivak et al. 2011).

21   Like most cultural ecologists, Nietschmann was an empiricist. One of his mantras was "keep collecting data!" He was also a militant of a sort, one who struggled alongside the Miskito during their war against the Sandinistas. Thus we might categorize him a militant empiricist. Yet I hesitate to do so. Nietschmann was the sort of Sauerian geographer who would have rebelled against the Bowman expeditions and their politics (see Wainwright 2008, chapter 6).

22   Partly for this reason, I (like Ismail and Spivak) travel internationally a great deal. On the relationship between travel, reflection, and "transcritique," see Karatani (2005, Introduction *passim*) and Karatani and Wainwright (2012, pp. 35–36).

23   Changing the disciplinary habits of geography has been on the agenda at least since the 1970s. In his now-classic statement on "revolutionary and counter- revolutionary theory" (1973), David Harvey writes:

The division of knowledge allows the body politic to divide and rule as far as the application of knowledge is concerned. It also renders much of the academic community impotent, for it traps us into thinking that we can understand reality only through a synthesis of what each discipline has to say about its particular segment and we quickly shrink away from what is so clearly an impossible task. [...] Reality has, therefore, to be approached directly rather than through the formulations of academic disciplines. We have to think in non- or meta-disciplinary terms if we are to think academically about our problems at all. (pp. 148–149)

I applaud Harvey's argument to this point, but disagree with his subsequent claim that "Geography has less of a problem than most [disciplines] in this regard since most geographers fortunately have little idea as to what geography is and are forced to make heavy use of other disciplines" (Ibid.). This is a non sequitur. On geography and "committed explanation," see Gregory (1978, chapter 5).

24   The last expression comes from a passage where Spivak writes apropos Edward Said: "Edward would ask, 'Gayatri, what do you do when you go to those villages?' I would give the usual answer, 'Hang out' (*Mitwegsein*, suspend previous training in order to train yourself, you know). The answer was not satisfactory" (2005, p. 524). Elsewhere Spivak explains that she uses the term "fieldwork" for this practice because "learning from below" sounds "too pious" (Spivak and Sharpe 2002, p. 620).

25   Although I have been to Oaxaca four times and have met with some of the leaders from these indigenous communities, I emphasize that my critique is legitimated neither by expertise nor by empiricism. Of course, my travel to Oaxaca enabled and informs my analysis. I see this as an attempt, to repeat Spivak's formula, at the inscription of fieldwork without transcoding (see Spivak and Sharpe 2002); compare Wainwright (2008, chapter 5).

26   I hope that someone else will take up this task.

27   Each text was originally written in Spanish; English translations have circulated.

28   For two competing versions of this story, compare Cruz (2011) and Herlihy (in Steinberg et al. 2011). Strangely, Herlihy's 2011 reply to the critical essays in *Political geography* makes no reference to Cruz's accompanying piece, even as Herlihy refers to Bryan's critique ~10 times.

29   I had the privilege of being present and speaking at this meeting.

DOI: 10.1057/9781137301758

6
Eight Theses on Geopiracy

Abstract: *The book's arguments crystallized in eight theses.*

Keywords: anti-disciplinary object; geopirates; imperial extraction; organic intellectuals; representation; world

Wainwright, Joel. *Geopiracy: Oaxaca, Militant Empiricism, and Geographical Thought.* New York: Palgrave Macmillan, 2013. DOI: 10.1057/9781137301758.

▶

How could we take the opportunity presented by these texts from Oaxaca to rethink the production of geographical knowledge? If not as empiricists, how might we know a place like Oaxaca? Can we, following Ismail, *abide* by Oaxaca? To address these questions I recapitulate and crystallize my arguments through a set of eight theses.

First thesis: we should aspire to but cannot abide by Oaxaca in Ismail's sense—and we will not be able to do so until we unlearn or destroy the implicit empiricism of the discipline. This means, to reiterate, that we must overcome the structural disciplinary conception whereby geographers serve the community (or the powerful) by going from "here" (the United States, the academy) to "there" (the field, Oaxaca) to gather empirical material to produce the truth.[1]

The philosophical materials necessary for overcoming this conception exist, but are covered in dust. In his very first prison notebook (Q1§105; 1992, p. 194), Antonio Gramsci asks, "Can modern thought go beyond empiricism-pragmatism and become widespread in America without a Hegelian phase?" I hope the answer is yes, because empiricism-pragmatism remains dominant and I cannot see a Hegelian phase coming soon. Yet I am not one to question Gramsci's hunches. Perhaps to help spur the "Hegelian phase" we should read Hegel's (1830) scintillating analysis of empiricism, beginning with his definition:

> In Empiricism lies the great principle that *whatever is true must be in the actual world and present to sensation.* [... T]he main lesson of Empiricism is that man must see for himself and feel that he is present in every fact of knowledge which he has to accept. (§38, my italics)

Hegel's critique follows:

> [T]here is a fundamental delusion in all scientific empiricism. It employs the metaphysical categories of matter, force, those of one, many, generality, infinity, etc.; following the clue given by these categories it proceeds to draw conclusions, and in so doing presupposes and applies the syllogistic form. And *all the while it is unaware that it contains metaphysics in wielding which, it makes use of those categories and their combinations in a style utterly thoughtless and uncritical.* (Ibid., my italics)

The problem with empiricism is one that it cannot sense and therefore cannot know: that in attempting to overcome metaphysics through the privileging of experience of "the actual world[,] present to sensation," empiricism exposes its lack of awareness of its own metaphysical bases and thus falls into a thoughtless and uncritical style.

DOI: 10.1057/9781137301758

Can we think in a better style? To rephrase Gramsci: could Anglo-American geographical thought transcend empiricism-pragmatism? Can we overcome militant empiricism? Can we abide by the world? *These are four iterations of the same question.*

Why do we geographers have such difficulties facing up to the myriad ways that our discipline remains implicated in the work of empire? The answer to this question is my *second thesis*: our underlying limitation comes from a traditional commitment to empiricism in service to the state/military. The extent of direct involvement in geographical scholarship waxes and wanes, but the tide of state/military support returns with celestial reliability. Even that research which is not directly funded by the state typically relies on state institutions, emulates the expeditionary model, and seeks to be justified by promises of "usefulness to public policy."[2] Much as we may deny it, today many geographers are executing their own expeditions in the spirit of Isaiah Bowman, adapting his empiricist epistemology to the times and keeping our conscience clear by inserting the adjective "critical" before "geography," "ethnography," "GIS," and so on.

Like the nascent return of piracy on the transcontinental shipping lanes, the emergence of geopiracy harks to an earlier era—the time of Bowman and geography militant—and not without a touch of nostalgia. The parallel goes farther. Whereas the Somali pirates' deft interceptions of oil tankers reveal intense fractures and inequalities in global political economy (see Samatar et al. 2010),[3] so do the geopirates' expeditions, for they constitute attempts by one particular state to make the entire world known, secure, and open for business. The contrast, of course, reveals the massive inequalities in the global command of geographical data collection and analysis: one sort of "pirate" relies upon hard-won existential knowledge of the sea, the other the calculative power of GPS satellites and expensive GIS datasets. These inequalities are obviously productive (in capitalism's space economy), but they are also strategic, insofar as the US military recognizes that hegemony requires the maintenance of geospatial capacities over its competitors (particularly China). A comparison of the "geopirates" with "real" pirates reveals other parallels. Both groups operate in spaces that are strategic yet on the margins of US hegemony (of course, the "geopirates" seek to bolster US power—the "real" pirates are more ambivalent). And both are motivated by material interests, from which ideology is inseparable.[4]

Third thesis: the geopirates sailing the high seas of good intentions do not fly the Jolly Roger, but flags of militant empiricism and liberal

DOI: 10.1057/9781137301758

intervention. They are no less easily recognized. In lieu of plain black background we can discern the vita of an ambitious "research man" (Heidegger 1938, p. 125); in place of the skull and crossbones, the dollar signs of state/military support. Since this may seem harsh, let me repeat that our criticism should not overemphasize individual personalities. Consider, for instance, the two key figures who inspired the letters from Oaxaca: Professors Dobson and Demarest. We could heap scorn on them but it would mean nothing. Knowingly or not, they are the intellectual bearers of the interests of a particular social group: "enlightened" liberal managers of the contemporary US capitalist empire. Their work may not be intellectually interesting but it advances definite interests. They are (to use Gramsci's term) *organic intellectuals* of a social group that seeks to militarize geographical thought.[5] This is why objecting to the Oaxaca controversy as the result of practical-ethical errors by individual geographers is necessary but wholly inadequate. They are not so much errors as effects.

This leads to the *fourth thesis*. So long as there are powerful social groups who see in geography a tool for correcting the decline of US hegemony, the discipline's latent empiricism will be at risk of being weaponized. As we have seen, the emergence of the Bowman expeditions can be accounted for by the need for the US state/military to grasp the entire world qua "human terrain." This apparent need is an effect of the eroding capacity of the US state to manage and sustain its global hegemony via US-centered "soft power" (ideology plus market access) while disciplining anti-systemic groups with threats of counter-attack (anchored in overwhelming air power). As Afghanistan, Iraq, and Libya amply demonstrate—consider how quickly US/NATO forces dominated the airspace in each case, yet how completely the US failed to "win" these wars, i.e., to win a meaningful hegemony—it is not enough for the US military to control the Earth's skies and inner orbit. The Petraeus counterinsurgency strategy, which grew out of the rubble of Afghanistan and Iraq, has only two main branches: more boots-on-the-ground ("surge"), and superior geographical-anthropological-political intelligence ("human terrain").[6] This dual strategy, born of specific geographical-political-economic circumstances, solicits geographers. Thus Professor Dobson is correct in recognizing that the decline of US hegemony enhances the strategic value of geographical thinking to the state/military. What is in play is our response to this evaluation. It is often said that "war is God's way of teaching geography," but I think it fairer to conclude that *war is*

DOI: 10.1057/9781137301758

a geographer's excuse for "playing God." We must cultivate the courage to refuse the excuse.

Thesis five: the problem is not strictly political-economic-military. The metaphysical roots of the enrollment of geography into military thinking grow in deep soil. If we excavate carefully we can discern the influence of Hume, Locke, and Descartes—and even, farther down, close to the bedrock, traces of the 1492 encounter (Dussel 1993). *It is in this geological sense that the Zapotec claim that the Bowman expeditions constitute "geopiracy" is best appreciated.* For if the expeditions were no mere misunderstanding or ethical lapse, but a more general assault by "parasites upon our region,"[7] then we can appreciate that geopiracy names a *process of imperial extraction* that is not temporally limited to the present, not ethically limited to the acts of discrete individuals, nor a simple matter of winning the consent of subjects. Rather it reflects the afterlife of the Colombian encounter and its production of a divided world—the same world that empirical geographers take unproblematically as the object of analysis.

Thesis six: being in the world solicits geographical thought, makes it inevitable everywhere. The world is a peculiar anti-disciplinary object. It is open to the existence of geographical thought, yet resists it. From this vantage the critique of "geopiracy" is at once *de Oaxaca* (i.e., reflects Oaxacan realities but also the creative-poetical naming of the matter) but also reflects an absolutely general resistance by world, or planetarity, to empirical closure. To cite the lapidary words of Gerry Kearns (2010, p. 201): "the world is not only to be apprehended through force." On the contrary. Those ways of grasping the world that require power are always marked by fundamental limits, for, as Kant (1795, p. 155) put it, "The possession of power inevitably corrupts the free judgment of reason."

In his essay on geography Conrad (1924) suggests that the "earth is a stage, and though it may be an advantage, even to the right comprehension of the play, to know its exact configuration, it is always the drama of human endeavor that will be the thing, with a ruling passion expressed by outward action marching perhaps blindly...." Perhaps in this spirit we could say that our task is to celebrate the stage and learn to appreciate the drama—without insisting upon total mastery of "its exact configuration," to allow our "outward action" to march less blindly than when we are led to statist-military ends. Needed: a coherent conception of the world *without "human terrain."*[8]

DOI: 10.1057/9781137301758

This leads to *thesis seven*: our confrontation with geopiracy calls for
a revolutionizing geographical questioning, an event that today cannot
be separated from the challenge laid down by postcolonial scholars
such as Spivak and Ismail. Gayatri Spivak concludes "Can the subaltern
speak?" by reminding us that "representation has not withered away"
(1988, p. 283). This is still true, and we could interpret geopiracy as an
interested repression of this truth. Geographers who deny their interested
participation in the production of representations of the world are
either too impatient to dwell on their work or lying to themselves; in
either event, geopiracy is a likely result. This does not mean, of course,
that representation is simply bad; indeed it is necessary—but so too, for
geographical thought as much as for political life, is *critique*. In politics,
representation is inevitable: the question is usually posed as: "Who will
represent our demands?" (This is representation as proxy, *vertreten*).[9]
In the empiricist social sciences, that question is always decided in
advance. The subject (ethnographer, geographer) will represent the
object (people qua ethnos, world). This is representation as portrait
(*darstellen*). There is no contradiction between openly avowing, "yes, of
course, representation is inevitable in political life," while condemning
the dynamic of empiricist interpretation, i.e., that the active subject
speaks for the object. The challenge is to be rigorously critical about such
epistemic violence. Militant empiricism fails here; hence its appeal—to
liberals like Dobson—will persist. For those of us who wish to abide by
Oaxaca it is worth remembering that to confront the subaltern "is not
to represent (*vertreten*) but to learn to represent (*darstellen*) ourselves"
(Spivak 1988, p. 259).

I conclude with *thesis eight*: the Oaxaca controversy passed quickly,
which is just as well, but the task of thoughtful questioning remains:
questioning our disciplinary habits of mind. The controversy should
provide an occasion for careful reflection upon the responsibility of geographical knowledge. Above and prior to questions about research ethics
or disciplinary regulation of human subjects, we need to deepen our
commitment to questioning the conditions of possibility for responsibility. Discussion of "research ethics" always arrive too late: our alpha and
omega is worldly critique. This is why we must extend the postcolonial
analysis of Western metaphysics and ask how geography's core qualities—empiricism, the fieldwork romance, our partnership with the state/
military—have shaped the discipline in ways that remain fundamental.
The occasion should lead us to question what it would mean to respond

DOI: 10.1057/9781137301758

to the texts from Oaxaca, not as counterinsurgents, but as geographical thinkers who desire a world without geopiracy.

Notes

1 One underlying limitation is that geographical research "is not done on people who can call the cops" (Nietschmann 2001, p. 183). Rather it is conducted "on the poor and unknown" (Ibid.). Studying and criticizing relatively rich, and powerful people is only one aspect of the task at hand.
2 Harvey (1974) remains the best critique of the romance of pragmatism in geography. Its opening line still stings: "General Pinochet is a geographer by training, and by all accounts he is successfully putting geography into public policy" (p. 18).
3 On piracy and oil tankers, see also French and Chambers (2010). Remember that our planet is imperiled not by the pirates but by the *successful* delivery and consumption of the fuel on the oil tankers they interrupt (IPCC 2007). The US military, which leads the international coalition to defeat oil-corridor piracy, is the largest consumer of energy in the US (Karbuz 2011). "Energy consumed per active duty military and civilian personal is 35 percent higher than the US energy consumption per capita, which is amongst the highest in the world. While consuming that amount of energy, DoD emitted 73 million metric tons of CO_2, corresponding to over 4 percent of the total emissions in USA" (Ibid.).
4 See Samatar, Lindberg, and Mahayni (2010) on the political economy of contemporary piracy. They contend that "modern piracy is partly induced by poverty, unemployment and the temptations brought about by the proximity of wealth and poverty," and that "resource robberies by states and multinational companies create unbearable living conditions for local populations, which compel the latter to resist these predations" (p. 1379). Surely this is true for their case, Somali piracy. Perhaps geopiracy may be seen as yet another form of "these predations."
5 Gramsci writes (Q12§11 1971, p. 5): "Every social group, coming into existence on the original terrain of an essential function in the world of economic production, creates together with itself, organically, one or more strata [*ceti*] of intellectuals which give it homogeneity and an awareness of its own function not only in the economic but also in the social and political fields." Dobson and Demarest, no less than the anthropologist Montgomery McFate, are representatives of such a *ceti*. It would be useful to have a Gramscian study of this group, which would, I hypothesize, demonstrate that it finds its coherence in the wake of the US state/military reaction to 9/11. For instance,

DOI: 10.1057/9781137301758

Dobson's views on the value of geography for the US state/military seem to have crystallized after September 2001—around the time when he left he left his position as a contract researcher at Oak Ridge National Laboratories, a US military research center, for the University of Kansas. While such individual pathways may be of interest, the general point I stress is that the political-economic conditions of the post-9/11 period created new opportunities for those opportunistic intellectuals, such as Dobson, who wished to raise their discipline's profile by accommodating state/military power. As Dobson explains (2007): "During my 26-year career at Oak Ridge National Laboratory and 6 years at the University of Kansas, I have participated in scores of meetings among insiders who provide decision support to foreign policy makers and military strategists. For about 2 years [i.e., 2004-06, presumably] I've sensed an historic opening for the restoration of geography. Many insiders now openly admit that geographic understanding is sorely missing from their deliberations, though only a few know to call it geography."

6   These are hardly new ideas. They bear obvious parallels to British colonial police tactics and reflect the limits of the Powell Doctrine in an era of imperial transition.

7   An evocative phrase from the July 24, 2011 meeting in Yagila, Oaxaca.

8   "Human terrain" is one species of a genus of concepts that have deep roots in cultural ecology. It is thus especially noteworthy that those accused of "geopiracy" by the indigenous communities in Oaxaca have wrapped themselves in the cloak of Carl Sauer, a father of geographical cultural ecology. They cite their ties to his legacy and reminding us that he too received military funding. But we should remember that Sauer's 1925 paper on "The Morphology of Landscape"—the textual "origin" of the concept of "cultural landscape," recently conceptualized for the Army as "human terrain"—represents an attempt to sketch a critical phenomenology of the world: "the task of geography," writes Sauer, should be "conceived as the establishment of a critical system which embraces the phenomenology of landscape" (1925, p. 25). Viewed retrospectively we should recognize that Sauer's phenomenology was insufficiently critical and only too amenable to militant empiricism. A contemporary of Sauer's, Martin Heidegger, comments that a "remodeling of the sciences [...] can only be accomplished [...] from a turning around of the attitude of knowing before all science. This turning around is only created through a long and unswerving execution of a revolutionizing questioning" (2009 [1934], §4, p. 7). The execution of Sauer's "task" in postcolonial terms, if such a thing is possible, would not establish "a critical system," but rather an anti-systemic critique: a conception of the world born of revolutionizing questioning.

9   On the politics of representation, compare Marx (1852), Spivak (1988), and Karatani (2012, chapter 1).

DOI: 10.1057/9781137301758

References

AAA (2007) American Anthropological Association Executive Board Statement on the Human Terrain System Project, October 31, 2007. Available at: http://www.aaanet.org/issues/policy-advocacy/Statement-on-HTS.cfm.

AAA (2009) AAA Commission on the Engagement of Anthropology with the US Security and Intelligence Communities. Final Report on the US Army Human Terrain System Proof of Concept Program, October 14, 2009. Available at: http://www.aaanet.org/cmtes/commissions/ceaussic/upload/ceaussic_hts_final_report.pdf.

AAA (2011) Commission on the Engagement of Anthropology with the US Security and Intelligence Communities. Available at: http://www.aaanet.org/cmtes/commissions/CEAUSSIC/index.cfm.

AAG Council (2009) Statement on professional ethics: revised November 1, 2009. Available at: http://www.aag.org/cs/resolutions/ethics.

AAUP (2000) Institutional Review Boards and social science research. Available at: http://www.aaup.org/AAUP/comm/rep/A/protecting.htm.

Agamben G (2005) *State of exception*. Chicago: University of Chicago.

Agnew J (2009) Common sense versus good sense. *AAG newsletter* 44(6), 3.

Agnew J (2010) Ethics or militarism? The role of the AAG in what was originally a dispute over informed consent. *Political geography* 29(8), 422–423.

94   *References*

AGS (2009) The American Geographical Society's Bowman Expeditions seek to improve geographic understanding at home and abroad: Spotlight on México Indígena (press release). Available at: http://web. ku.edu/wmexind/bowmanPRespangles.pdf.

Althusser L (2005 [1965]) *For Marx.* New York: Verso.

Appendini K (2001) Land regularization and conflict resolution: The case of Mexico. Food and Agriculture Organization of the United Nations. Accessed at: http://faorlc.cgnet.com/es/desarrollo/interag/pdf/mexico.pdf.

Barnes T (2006) Geographical intelligence: American geographers and research and analysis in the Office of Strategic Services 1941–1945. *Journal of historical geography* 32, 149–168.

Barnes T (2008) Geography's underworld: the military-industrial complex, mathematical modeling and the quantitative revolution. *Geoforum* 39, 3–16.

Barnes T and Farish M (2006) Between regions: science, militarism, and American geography from world war to cold war. *Annals of the Association of American Geographers* 96(4), 807–826.

Bateson G (1972 [1969]) The double bind. In *Steps toward an ecology of mind.* New York: Chandler Publishing.

Batson D (2008) Registering the human terrain: a valuation of cadastre. Washington, DC: NDIC Press, National Defense Intelligence College. Available at: http://handle.dtic.mil/100.2/ADA485498.

Blaut J (1969) Jingo geography: part I. *Antipode* 1(1), 10–13.

Bok D (1982) *Beyond the ivory tower: social responsibilities of the modern university.* Cambridge: Harvard University Press.

Bowman I (1930) *Geography in relation to the social sciences.* New York: Scribner.

Boyce G and Cash, C forthcoming (2013) Geography, counterinsurgency, and the "G-Bomb": the Case of *México Indígena.* In W Munger and K Williams (eds.) *Life during wartime: counterinsurgency and strategic studies.* Oakland: AK Press.

Bryan J (2010) Force multipliers: geography, militarism, and the Bowman Expeditions. *Political geography* 29(8), 414–416.

Bryan J and Wainwright J (2009) Letter to the Association of American Geographers. Accessed at: http://academic.evergreen.edu/g/grossmaz/HerlihyLetterSign.pdf.

Bumiller E (2012) After war room, heading Ivy League classroom. *New York times* May 6, 2012. Accessed at: http://www.nytimes.

DOI: 10.1057/9781137301758

com/2012/05/07/us/retired-military-officers-teaching-at-ivy-league-schools.html?_r=2&pagewanted=1.

Chapin M and Threlkeld B (2001) *Indigenous landscapes: a study in ethnocartography.* Arlington, VA: Center for the Support of Native Lands.

Chapin M, Lamb Z, and Threlkeld B (2005) Mapping indigenous lands. *Annual review of anthropology* 34, 619–638.

Chatterjee P (2001[1986]) *Nationalist thought and the colonial world: a derivative discourse.* Minneapolis: University of Minnesota Press.

Cloud J (2002) American cartographic transformations during the Cold War. *Cartography and geographic information science* 29, 261–282.

Cockburn P (2011) Wars without victory; America without influence. *Counterpunch* December 12, 2011. Accessed at: http://www.counterpunch.org/2011/12/12/wars-without-victory-america-without-influence/.

Cockburn A and St. Clair J (2004) *Imperial crusades: Iraq, Afghanistan and Yugoslavia.* New York: Verso.

Colten C (2009) AAG Council Meeting Minutes, March 20–22, 2009. *AAG newsletter* July/August, 20–25.

Conrad J (1972 [1924]) Geography and some explorers. In *Last essays,* R Curle (ed.) London: Dent & Sons, 1–21.

Craib R (2004) *Cartographic Mexico: a history of state fixations and fugitive landscapes.* Durham, NC: Duke University.

Crampton J (2003) Cartographic rationality and the politics of geosurveillance and security. *Cartography & GIS* 30(2), 131–144.

Crampton J (2006) The cartographic calculation of space: race mapping and the Balkans at the Paris Peace Conference of 1919. *Social and cultural geography* 7(5), 731–752.

Crampton J (2007) The biopolitical justification for geosurveillance. *Geographical review* 97(3), 389–403.

Crampton J (2008) The role of geosurveillance and security in the politics of fear. In D Sui (ed.) *Geospatial technologies and homeland security: research frontiers and challenges.* Springer-Verlag, pp. 283–300.

Crampton J and Barnes T (2011) Mapping intelligence. American geographers and the Office of Strategic Services and GHQ/SCAP (Tokyo). In S Kirsch and C Flint (eds.) *Reconstructing conflict.* London: Ashgate, pp. 227–251.

Cruz M (2010) A living space: the relationship between land and property in the community. *Political geography* 29(8), 420–421.

DOI: 10.1057/9781137301758

Deininger K, Lavadenz I, Bresciani F, and Diaz M (2001) *Mexico's "second agrarian reform": implementation and impact.* Washington DC: The World Bank.

de Ita A (2006) Land Concentration in Mexico after PROCEDE. In P Rosset and R Patel (eds.) *Promised land: competing visions of agrarian reform.* Oakland, CA: Food First, pp. 148–164.

de Janvry A, de Anda Gordillo G, and Sadoulet E (1997) *Mexico's second agrarian reform: household and community responses, 1990–1994.* La Jolla, CA: Center for US–Mexican Studies, University of California–San Diego.

Demarest G (1995) *The overlap of military and police in Latin America.* Fort Leavenworth, Kansas: FMSO.

Demarest G (1998) *Geoproperty: foreign affairs, national security, and property rights.* London: Frank Cass.

Demarest G (2003) *Mapping Colombia: the correlation between land data and strategy.* Carlisle, PA: US Army War College. Available at: http://www.strategicstudiesinstitute.army.mil/pubs/display.cfm?pubid¼412.

Demarest G (2009 [2007]) *Property and peace: insurgency, strategy and the Statute of Frauds* Fort Leavenworth, KS: FMSO. Accessed at: http://fmso.leavenworth.army.mil/documents/Property-and-Peace.pdf.

Derrida J (1978 [1966]) Structure, sign and play in the discourse of the human sciences. *Writing and difference.* Alan Bass (trans.). Chicago: University of Chicago, 278–294.

Derrida J (1976 [1967]) *Of grammatology.* G Spivak (trans.). Baltimore: Johns Hopkins.

Derrida J (1982 [1968]) *Différance.* In *Margins of philosophy.* Chicago: University of Chicago Press, pp. 1–27.

Derrida J (1990) Force of law: the "mystical foundation of authority." *Cardozo law review* 11(5–6), 920–1045.

Deutscher I (1967[1949]) *Stalin: a political biography.* Oxford: University of Oxford.

Dobson J (2005a) Foreign intelligence is geography. *Ubique* XXV(1), 1–2.

Dobson J (2005b) The Dawning of the G-bomb. *Directions magazine.* Accessible at: http://www.directionsmag.com/articles/the-dawning-of-the-g-bomb/123453.

Dobson J (2006a) AGS conducts fieldwork in Mexico. *Ubique* XXVII(1), 1–3.

Dobson J (2006b) Fort Leavenworth hosts AGS council. *Ubique* XXVII(3), 1–3.

DOI: 10.1057/9781137301758

Dobson J (2006c) Restoring geography in America. *Ubique* XXVII(3), 1–2.

Dobson J (2007) Bring geography back! *ArcNews online*. Available at: http://www.esri.com/news/arcnews/spring07articles/bring-back-geography-1of2.html.

Dobson J (2009) Let the indigenous people of Oaxaca speak for themselves. February 5, 2009. *Ubique* XXIX (1), 1–11. Accessed at: http://academic.evergreen.edu/g/grossmaz/BowmanEthics.doc.

Dobson J (2010) Geography: use it or lose it. Presentation to the US State Department, Washington, DC. May 25, 2010. Accessed at: http://www.state.gov/e/stas/series/154218.htm.

Dobson J (2011) Through the macroscope: geography's view of the world. Presentation at Ohio State University, November 17, 2011.

Dobson J and Fisher P (2003) Geoslavery. *IEEE technology and society magazine* 22(1), 47–52.

Driver F (2001) *Geography militant: cultures of exploration and empire*. Oxford: Wiley-Blackwell.

Denham D and CASA collective (2008) *Teaching rebellion: stories from the grassroots mobilization in Oaxaca*. Oakland, CA: PM.

Deutscher I (1967 [1949]) *Stalin: a political biography*. Oxford: Oxford University.

Dussel E (1993) Eurocentrism and modernity. *Boundary* 2, 20(3), 65–76.

ETC Group (2010) *Geopiracy: the case against geoengineering*. Accessed at: http://www.etcgroup.org/upload/publication/pdf_file/ETC_geopiracy_4web.pdf.

ETC Group (2011) Earth grab: geopiracy, the new biomasters and capturing climate genes. Cape Town: Pambazuka.

Foucault M (1994 [1966]) *The order of things: an archaeology of the human sciences*. New York: Vintage.

Foucault M (2008 [1978–1979]) *The birth of biopolitics*. New York: Picador.

French P and Chambers S (2010) *Oil on water: tankers, pirates, and the rise of China*. London: Zed.

Fried G (2000) *Heidegger's polemos: from being to politics*. New Haven, CT: Yale University.

Gajilan A (2004) Entrepreneurs in Iraq tangle in US red tape. *Fortune small business*. November 1, 2004. Accessed at: http://money.cnn.com/magazines/fsb/fsb_archive/2004/11/01/8190934/index.htm.

Gibler J (2009) *Mexico unconquered: chronicles of power and revolt*. New York: City Lights.

DOI: 10.1057/9781137301758

Gillespie T, Agnew J, Mariano E, Mossle S, Jones N, Braughton M, and Gonzalez J (2009) Finding Bin Laden: an application of biogeographic theories and satellite imagery. *California center for population research on-line paper* CCPR–002–09.

Gidwani V (2008) *Capital, interrupted: agrarian development and the politics of work in India.* Minneapolis: University of Minnesota.

González R (2007) Towards mercenary anthropology? *Anthropology today* 23(3), 14–19.

González R (2009) *American counterinsurgency: human science and the human terrain.* Chicago: Prickly Paradigm Press.

Goodchild M (2010) Public commentary at the University of Arizona School of Geography and Development. March 5, 2010.

Gramsci A (1971) *Selections from the prison notebooks.* Q Hoare and G Smith (eds and trans.). New York: International.

Gramsci A (1992) *Prison notebooks volume I.* Buttigieg J (ed. and trans.). New York: Columbia University.

Grandin G (2005) *Empire's workshop: Latin America, the United States, and the rise of the new empire.* New York: Metropolitan.

Grant S (2011) Selective overview of US Army human geography research programs. January 27, 2011. Accessed at: http://www.wbresearch.com/uploadedfiles/Events/UK/2012/10980_006/Info_for_Attendees/presentations/1100%20Grant.pdf.

Gregory D (1978) *Ideology, science and human geography.* London: Hutchinson.

Gregory D (2004) *The colonial present.* Oxford: Blackwell.

Gregory D (2008) "The rush to the intimate": counter-insurgency and the cultural turn. *Radical philosophy* 150, 8–23.

Gregory D (2011) From a view to a kill: drones and late modern war. *Theory culture & society* 28(7–8), 188–215.

Gusterson H (2003) Anthropology and the military: 1968, 2003, and beyond? *Anthropology today* 19(3), 25–26.

Hartshorne R (1939) *The nature of geography.* Lancaster, PA: Association of American Geographers.

Harvey D (1969) *Explanation in geography.* London: Edward Arnold.

Harvey D (2009 [1973]) *Social justice and the city.* Athens, GA: University of Georgia.

Harvey D (1974) What kind of geography for what kind of public policy? *Transactions of the Institute of British Geographers* 63, 18–24.

Harvey D (2003) *The new imperialism.* Oxford: Oxford University.

DOI: 10.1057/9781137301758

Heidegger M (1977 [1938]) The age of the world picture. In *The question concerning technology and other essays*. W. Lovitt., trans. and ed. New York: Harper, pp. 115–154.

Heidegger M (2009 [1934]) *Logic as the question concerning the essence of language*. Albany: SUNY Press.

Heidegger M (2010 [1927]) *Being and time*. J. Stambaugh, trans.; revised by D. Schmidt. Albany, NY: SUNY Press.

Hegel G W F (2008 [1892]) *Hegel's Logic: part one of the encyclopaedia of the philosophical sciences*. Oxford: Oxford University Press.

Heller-Roazen D (2009) *The enemy of all: piracy and the law of nations*. New York: Zone.

Herlihy P (2010a) Panelist presentation, research ethics for Latin Americanist Geographers, annual meeting of the AAG. April 15, 2010. Washington DC.

Herlihy P (2010b) Self-appointed gatekeepers attack the American Geographical Society's first Bowman Expedition. *Political geography* 29(8), 417–419.

Herlihy P, Dobson J, Aguilar Robledo M, Smith D, Kelly J, and Ramos Viera A (2008) A digital geography of indigenous Mexico: prototype for the American Geographical Society's Bowman expeditions. *The geographical review* 98(3), 395–415.

Herlihy P, Dobson J, Aguilar Robledo M, Smith D, Kelly J, Ramos Viera A, and Hilburn A (2008a) *México Indígena*. Final Report to FMSO (short version for web): The AGS Bowman Expeditions prototype: digital geography of indigenous Mexico. Unpublished manuscript. Available at: web.ku.edu/~mexind/FMSO_Final_Report_2008_web_version.pdf.

Herlihy P and Knapp G (2003) Maps of, by, and for the peoples of Latin America. *Human organization* 62(4), 303–314.

Herlihy P, Smith D, Kelly J, and Dobson J (2006) Executive summary of *México Indígena*: Mexican open-source Geographic Information Systems (GIS) project final report, year one. Unpublished manuscript prepared by the Radiance Research Team for the FMSO and the AGS.

Hernández R and Montaño Mendoza B[the civil authorities of Tiltepec] (2009) Statement on behalf of the community of San Miguel Tiltepec, Ixtlán de Juárez. March 17, 2009. Available at: http://academic.evergreen.edu/g/grossmaz/bowman.html.

Heyman R (2009) People *can:* the geographer as anti-expert. *Acme* 9(3), 301–328.

DOI: 10.1057/9781137301758

Horkheimer T and Adorno T (2002 [1947]) *The dialectic of Enlightenment.* E Jephcott (trans.). Stanford: Stanford University Press.

Intergovernmental Panel on Climate Change (IPCC) (2007) Climate Change 2007: Synthesis report: Summary for policymakers. Accessed July 6, 2009, at http://www.ipcc.ch/pdf/assessment-report/ar4/syr/ar4_syr_spm.pdf.

International Human Rights and Conflict Resolution Clinic (Stanford Law School) and Global Justice Clinic (NYU School of Law) (2012) *Living under Drones: Death, Injury, and Trauma to Civilians from us Drone Practices in Pakistan.* Accessed at http://www.law.stanford.edu/sites/default/files/organization/149662/doc/slspublic/Stanford_NYU_LIVING_UNDER_DRONES.pdf.

IPSG (2009a) Letter to the AAG from the Indigenous Peoples Specialty Group. April 14, 2009. Available at: http://academic.evergreen.edu/g/grossmaz/IPSGletterAAGboard.pdf.

IPSG (2009b) Letter to San Miguel Tiltepec, Oaxaca, from the Indigenous Peoples Specialty Group. April 14, 2009. Available at: http://academic.evergreen.edu/g/grossmaz/IPSGLetterSanMiguelTiltepec.pdf.

IPSG (2010) Indigenous Peoples Specialty Group's declaration of key questions about research ethics with indigenous communities. Available at: http://academic.evergreen.edu/g/grossmaz/IPSGResearchEthicsFinal.pdf.

Ismail Q (2005) *Abiding by Sri Lanka: peace, place, and postcoloniality.* Minneapolis: University of Minnesota Press.

Jan N (2011) Notes on metacoloniality. Unpublished manuscript.

Jazeel T (2011) Spatializing difference beyond cosmopolitanism: rethinking planetary futures. *Theory culture & society* 28(5), 75–97.

Kant I (1965 [1787]) *Critique of pure reason.* New York: Macmillan.

Kant I (2004 [1795]) Perpetual peace: a philosophical sketch. In H Reiss (ed.) *Kant: political writings.* Cambridge: Cambridge University, 93-130.

Karatani K (2008) Beyond capital-nation-state. *Rethinking Marxism* 20(4), 569–595.

Karatani K (2012) *History and repetition.* New York: Columbia University.

Karatani K and Wainwright J (2012) "Critique is impossible without moves": an interview with Kōjin Karatani. *Dialogues in human geography* 2(1), 30–52.

DOI: 10.1057/9781137301758

Karbuz S (2011) A look at US military energy consumption. Accessed at: http://oilprice.com/Energy/Energy-General/A-Look-At-US-Military-Energy-Consumption.html.

Katz C (1994) Playing the field: questions of fieldwork in geography. *The professional geographer* 46, 67–72.

Kearns G (2010) Geography, geopolitics and empire. *Transactions of the Institute of British Geographers* 35(2), 187–203.

Kelly J, Herlihy P, Smith D, and Ramos Viera A (2010) Indigenous territoriality at the end of the social property era in Mexico. *Journal of Latin American geography* 9(3), 161–181.

Kelly J, Jauregui B, Mitchell S, and Walton J (eds.) (2010) *Anthropology and global counterinsurgency*. Chicago: University of Chicago.

Kipp J, Grau L, Prinslow K, and Smith D (2006) The human terrain system: a CORDS for the 21st century. *Military review*, 8–15. Accessed at: http://www.army.mil/professionalWriting/volumes/volume4/december_2006/12_06_2.html.

Klein N (2004) Baghdad year zero: pillaging Iraq in pursuit of a neocon utopia. *Harper's magazine*. September 2004. Accessed at: http://harpers.org/archive/2004/09/0080197.

Louis R and Grossman Z (2009): see IPSG (2009).

Luttwak E (2007) Dead end: counterinsurgency warfare as military malpractice. *Harper's magazine*. February, 33–42.

Madson K (2008) Indigenous research, publishing, and intellectual property. *American Indian culture and research journal* 32(3), 89–105.

Marcus G (2009) CEAUSSIC: origin story and grand finale. Available at: http://blog.aaanet.org/2009/12/07/ceaussic-origin-story-and-grand-finale/.

Martin P (2008) Citizenship and the imperial city. *Antipode* 40(2), 221–225.

Martin P and Gática R (2008) An interview with Raúl Gática from the Popular Indigenous Council of Oaxaca–Ricardo Flores Magón (CIPO-RFM). *Antipode* 40(2), 211–215.

Martínez R (2008) La asemblea popular de los pueblos de Oaxaca (APPO). In *Los movimientos sociales de siglo XXI*. Caracas: Fundación editorial el perro y la rana, pp. 173–186.

Marx K (1852) The Eighteenth Brumaire of Louis Bonaparte. Accessed at: the Marx/Engels Internet Archive at http://www.marxists.org/archive/marx/works/1852/18th-brumaire/.

DOI: 10.1057/9781137301758

Marx K and Engels F (1848) *Manifesto of the communist party*. Accessed at: the Marx/Engels Internet Archive at http://www.marxists.org/archive/marx/works/1848/communist-manifesto/.

McFate M (2005) Anthropology and counterinsurgency: the strange story of their curious relationship. *Military review* (March–April), 24–38.

McFate M and Jackson A (2005) An organizational solution for DoD's cultural knowledge needs. *Military review* (July–August), 18–21.

McSweeney K (2010) Panelist presentation, "Research ethics for Latin Americanist Geographers," annual meeting of the AAG. April 15, 2010. Washington DC.

México Indígena (ca 2005) Problematic, focus, and objectives of the México Indígena Bowman Expedition prototype. Available at: http://web.ku.edu/~mexind/ags_problematic.htm.

México Indígena (July 2005–February 2008) Monthly reports to FMSO. Available at: web.ku.edu/~mexind/FMSO_WebReport.doc.

Moore S and Rivera M (2011) *Planetary loves: Spivak, postcoloniality, and theology*. New York: Fordham.

Morin K (2011) *Civic discipline: geography in America, 1860–1890*. London: Ashgate.

Municipal Authorities [Autoridades Municipales y Comisariados de Bienes Comunales de las comunidades de San Juan Tepanzacoalco, Santa María Zoogochi, Santa Cruz Yagavila, Santiago Teotlaxco y San Juan Yagila] (2011) Declaración Xidza sobre geopiratería. Unpublished manuscript. Yagila, Mexico, July 24, 2011.

Mutersbaugh T (2008) Oaxaca: terror and non-violent protest in a video age. *Antipode* 40(2), 205–210.

Mutersbaugh T (2009) Proposal for strengthening of geodata research protocols. Unpublished manuscript (internet post on Oaxaca studies listserv). January 28, 2009.

Mutersbaugh T (2010) Panelist presentation, research ethics for Latin Americanist Geographers, annual meeting of the AAG. April 15, 2010, Washington DC.

Nagar R (2002) Footloose researchers: "traveling" theories and the politics of transnational feminist praxis. *Gender, place and culture* 9, 179–186 .

Nagl J, Petraeus D, Amos J, and Sewall S (2007 [2006]) *The US Army/Marine Corps counterinsurgency manual*. Chicago: University of Chicago Press.

DOI: 10.1057/9781137301758

National Geospatial-Intelligence Agency (2012) Human geography provides context to GEOINT. *Pathfinder* 10(5), 13–15.

Network of Concerned Anthropologists (2009) *The counter-counterinsurgency manual, or, notes on demilitarizing American society.* Chicago: Prickly Paradigm.

Nietschmann B Q (1995) Defending the Miskito reefs with maps and GPS: mapping with sail, scuba, and satellite. *Cultural survival quarterly* 18(4), 34–37.

Nietschmann B Q (2001) The Nietschmann syllabus: a vision of the field. *The geographical review* 91(1–2), 175–184.

O'Laughlan J (2005) The war on terrorism, academic publication norms, and replication. *The professional geographer* 57(4), 588–591.

Ó Tuathail G (1996) *Critical geopolitics.* New York: Routledge.

Petraeus D (2006) Learning counterinsurgency: observations from soldiering in Iraq. *Military review,* 45–55.

Petraeus D (2007) Report on the status of Iraq: testimony before the US Congress. September 10, 2007. Accessed March 12, 2012, at: http://www.defense.gov/pubs/pdfs/Petraeus-Testimony20070910.pdf.

Petraeus D and Amos J (2006) Foreword. In *The US Army and Marine Corps counterinsurgency field manual.* Chicago: University of Chicago, xlv–xlvi.

Price D (2009) Faking scholarship: domestic propaganda and the republication of the *Counterinsugency field manual.* In NWA, *The counter-counterinsurgency manual* Chicago: Prickly Paradigm, 59–76.

Price D (2011) *Weaponizing anthropology.* Oakland, CA: Counterpunch/AK press.

Rich A (2011 [2001]) Credo of a passionate skeptic. *Los Angeles times.* March 11, 2001. Republished in *Monthly review* 64(2), 36–41.

Said E (1979) *Orientalism* New York: Pantheon.

Said E (1994) *Representations of the intellectual.* New York: Vintage.

Said E (2002 [1988]) Representing the colonized: anthropology's interlocutors. In *Reflections on exile.* Cambridge: Harvard University, 293–316.

Samatar A, Linberg M, and Mahaynia B (2010) The dialectics of piracy in Somalia: the rich versus the poor. *Third world quarterly* 31(8), 1377–1394.

Sanders M (2006) *Gayatri Chakravorty Spivak: live theory.* New York: Continuum.

DOI: 10.1057/9781137301758

Sauer C (1925) The morphology of landscape. *University of California publications in geography* 2(2), 19–53.

Sedillo S (2007) The road to hell. *El Enemigo Común*. November 26, 2007. Accessible at: http://elenemigocomun.net/1368/x/en.

Sedillo S (2009) Threat of genocide: US military mapping against Mexico's indigenous. *Left turn.* July 2009. Available at: http://www.leftturn.org/threat-of-genocide.

Sedillo S (2010) *The Demarest factor / El factor Demarest*. Manoveulta films. 60 minutes.

Sewall S (2006) A radical field manual. In *The US Army and Marine Corps counterinsurgency field manual*. Chicago: University of Chicago, xxi–xliii.

Sharp J and Dowler L (2011) Framing the field. In Del Casino V, M Thomas, P Cloke, and R Panelli (eds.) *A companion to social geographies*. Oxford: Blackwell, pp. 146–160.

Shaw I and Akhter A (2012) The unbearable humanness of drone warfare in FATA, Pakistan. *Antipode* doi: 10.1111/j. 1467–8330.2011.00940.x.

Smith D, Herlihy, P, Kelly, J, and Ramos Viera A (2009) The certification and privatization of indigenous lands in Mexico. *Journal of Latin American geography* 8(2), 175–207.

Smith N (1994) Geography, empire and social theory. *Progress in human geography* 18(4), 491–500.

Smith N (2003) *American empire: Roosevelt's geographer and the prelude to globalization*. Berkeley, CA: University of California.

Social Science Research Council (SSRC) (2008) The Minerva Controversy. Available at: http://essays.ssrc.org/minerva/.

Spivak G C (1999) *A critique of postcolonial reason*. Cambridge: Harvard University.

Spivak G C. (2003) *The death of a discipline*. New York: Columbia University.

Spivak G C (2005) Thinking about Edward Said: pages from a memoir. *Critical inquiry* 31, 519–525.

Spivak G C (2008) *Other Asias*. Oxford: Wiley-Blackwell.

Spivak G C (2010 [1988]) Can the subaltern speak? In R Morris (ed.) *Can the subaltern speak? Reflections on the history of an idea*. New York: Columbia.

Spivak G C (2010) Reading the global turn. *Frontier* 43(19), November 21–27.

Spivak G C (2011) Marx said the realm of freedom . . . *Frontier* 44(20), November 27–December 3.

DOI: 10.1057/9781137301758

Spivak G C (2012) *An aesthetic education in an age of globalization.* Boston: Harvard.

Spivak G C, Jones S, Keller C, Pui-Lan K, and Moore S (2011) Love: a conversation. In S Moore and M Rivera (eds.) *Planetary loves: Spivak, postcoloniality, and theology.* New York: Fordam, pp. 55–78.

Spivak G C and Sharpe J (2002) A conversation with Gayatri Chakravorty Spivak: politics and the imagination. *Signs* 28(2), 609–624.

Steinberg P, Bryan J, and Herlihy P (2011) Discussion: responses to the Bowman Expedition editorials. *Political geography* 30(2), 110.

Stephen L (2002) *Zapata lives!: histories and cultural politics in southern Mexico.* Berkeley: University of California Press.

Sui D (2007) Geospatial technologies for surveillance: tracking people and commodities in real-time. *Geographical review* 93(3), 3–9.

Sundberg J (2003) Masculinist epistemologies and the politics of fieldwork in Latin Americanist geography. *The professional geographer* 55(2), 181–191.

Swarr A L and Nagar, R (eds.) (2010) *Critical transnational feminist praxis.* New York: SUNY Press.

Twain, M (1917 [1901]) The Battle Hymn of the Republic (brought down to date). In *The complete works of Mark Twain*, vol. 20. New York: Harper Brothers, p. 465.

Unión de Organizaciones de la Sierra Juárez, Oaxaca (UNOSJO) (2009) Press bulletin from the Union of Organizations of the Sierra Juárez of Oaxaca (UNOSJO). Available at: http://www.grassrootsonline.org/news/blog/zapotec-indigenous-peoplemexico-demand-transparency-us-scholar.

Vogel J, Robles J, Gomides C, and Muniz C (2008) Geopiracy as an emerging issue in Intellectual Property Rights: the rationale for leadership by small states. *Tulane environmental law journal* 21, 391–406.

Wainwright J (2005) The geographies of political ecology: after Edward Said. *Environment and planning A* 37(6), 1033–1043.

Wainwright J (2008) *Decolonizing development: colonial power and the Maya.* Oxford: Blackwell Press.

Wainwright J and Bryan J (2009a) Cartography, territory, property: postcolonial reflections on indigenous counter-mapping in Nicaragua and Belize. *Cultural geographies* 16, 153–178.

Wainwright J and Bryan J (2009b) Open letter to geographers. Unpublished manuscript available at: http://academic.evergreen.

DOI: 10.1057/9781137301758

edu/g/grossmaz/Bryan%20%26%20Wainwright%208%20April%20
2009.pdf.

Wallace W and Hemenway R (2004) Memorandum of understanding between the University of Kansas and Ft Leavenworth. Unpublished manuscript.

Wallerstein I, et al. (1994) *Open the social sciences: report of the Gulbenkian Commission on the Restructuring of the Social Sciences.* Stanford, CA: Stanford University.

Wilshusen P (2010) The receiving end of reform: everyday responses to neoliberalisation in Southeastern Mexico. *Antipode* 42(3), 767–799.

Woodward R (2005) From military geography to militarism's geographies: disciplinary engagements with the geographies of militarism and military activities. *Progress in Human geography* 29(6), 718–740.

Woolhouse R (1990) *The empiricists.* Oxford: Oxford University.

Wright M (2008) In the name of democracy. *Antipode* 40(2), 200–204.

Žižek S (2004) What Rumsfeld doesn't know that he knows about Abu Ghraib. Available at: http://www.lacan.com/zizekrumsfeld.htm.

DOI: 10.1057/9781137301758

Index

DOI: 10.1057/9781137301758

DOI: 10.1057/9781137301758

Lightning Source UK Ltd.
Milton Keynes UK
UKOW041803020613

211625UK00001B/3/P